Flows and Chemical Reactions in Heterogeneous Mixtures

Flows and Chemical Reactions in Heterogeneous Mixtures

Roger Prud'homme

iSTE

WILEY

First published 2014 in Great Britain and the United States by ISTE Ltd and John Wiley & Sons, Inc.

ISTE Ltd
27-37 St George's Road
London SW19 4EU
UK

www.iste.co.uk

John Wiley & Sons, Inc.
111 River Street
Hoboken, NJ 07030
USA

www.wiley.com

Library of Congress Control Number: 2014950493

British Library Cataloguing-in-Publication Data
A CIP record for this book is available from the British Library
ISBN 978-1-84821-785-0

Contents

Preface

This book – a continuation of two previous publications, respectively dealing with *"Flows and chemical reactions"* [PRU 12] and *"Flows and chemical reactions in homogeneous mixtures"* [PRU 13] – is dedicated to *"Flows and chemical reactions in heterogeneous mixtures"*.[1]

Note that the concept of heterogeneity in fluids is entirely relative [DOR 89].[2] The heterogeneous media presented in this volume notably include interfaces and lines. They may be the site of radiative transfer. A separate chapter is entirely dedicated to each type of flow in this book.

---

1 Remember that the volume *General Equations* comprised three sections: 1. Fluid media with a single component, 2. Reactive mixtures, and 3. Interfaces and lines, and that the volume *Flows and Chemical Reactions in Homogeneous Mixtures* comprised: 1. Nozzle flows, 2. Chemical reactors, and 3. Laminar and turbulent flames.

2 In the publication cited here, the following remarks are made about the concept of heterogeneity:

– a fluid is said to be heterogeneous if it is possible to distinguish one particle from its neighbor *and* we choose to make that distinction;

– the heterogeneity may refer to various aspects: differences in phase, chemical species, velocity, temperature, etc.;

– the size of the particles in question and the diagnostic tools at our disposal are two crucial elements in shaping our ability to discern these differences.

Multiphase flows – whether they be gaseous flows with solid or liquid particles, liquid flows with gas or vapor bubbles, or solid/liquid particles within another phase – involve exchanges between the phases, chemical reactions, disintegration and clustering of particles, as well as the various interactions between these particles.

It is necessary to model these flows in order to discover, for example, what happens in the atmosphere, or predict the performances of a fuel injection engine or determine the risk of explosion at a grain silo.

The description of fluid–particle exchanges is a problem in itself, and requires modeling of the particle at the *mesoscopic scale*[3], to determine, the effects of friction, or of heat- and mass exchange (e.g. the regression rate of a drop of fuel in combustion, which would help calculate the necessary length for a rocket engine). We then need to transcribe these exchange laws to the space of multi-phase mixing and write the balance equations, before going on to solve them.

A variety of methods can be used to establish the balance equations on the basis of a mixture's properties on the molecular scale. In particular, we can cite the techniques consisting of using a probabilistic balance equation, similar to that used in Chapter 2 of Volume 2 [PRU 13], and applying it to the different quantities [BAR 73, KUE 73].

It is also possible to use Direct Numerical Simulation (DNS), or the Monte Carlo method.

To begin, in *Chapter 1*, we shall examine the question of the generation of specific two-phase media: dusty gases, mists, bubbly flows, etc. The production of vapor for combustion is a complex problem, with multiple steps

3 That is to say, intermediary between the microscopic and macroscopic scales.

grouped together into primary atomization and secondary atomization.

Chapter 2 is given over to examination at the *mesoscopic* scale: exchanges of momentum and heat between a particle and a fluid, with determination of the respective exchange coefficients. Thus, we specify the values of the phenomenological coefficients which come into play in Chapter 3, where the laws of linearized TIP[4] are applied to multiphase flows at the macroscopic scale. Then we deal with coupled problems, such as that of the combustion of a drop of fuel (already discussed, in certain specific conditions, in Appendix 3 of Volume 1 of this series [PRU 12] and in Chapter 3 of Volume 2 [PRU 13]).

In *Chapter 3*, we establish simplified *macroscopic* balance equations for mass, momentum and energy, in the case of particles of exactly the same size (i.e. a monodisperse suspension). The forms of the constitutive relations are deduced from the principles of TIP. These equations are applied to typical problems: propagation of sound, and nozzle flow of a dilute suspension of solid particles.

Next, in *Chapter 4*, we examine flows with evaporating droplets, by first establishing the balance equations and then applying them to the process of combustion. Balances are first established for droplets assuming they have a uniform internal temperature; then balance equations of the two-phase mixture; and gas phase equations are deduced by subtracting those of the droplets. Finally, the balance of entropy of the mixture can be written, and phenomenological relations are deduced. We then look at the case where the internal temperature of the droplets is not uniform by using

4 TIP: Thermodynamics of Irreversible Processes [GRO 69a].

a multilayer model. An example of application is the study of the oscillatory instability of a flame spray in the presence of thermo-acoustic coupling.

Radiative phenomena are presented in *Chapter 5*. Such phenomena are active, in particular, in the case of production of soot resulting from the combustion of hydrocarbons, but also in problems of re-entry into planetary atmospheres and the combustion of certain solid propellants. The subject is dealt with by modeling of the radiative properties and the establishment of macroscopic balance equations. The chapter goes on to discuss a number of applications.

In the *Appendix*, the concept of the Hopf bifurcation is introduced.

Note that the type of equations obtained and the methods used are, of course, subject to limitations. As long as the flow regime is continuous and regular (similar to a laminar regime in a homogeneous fluid), the balance equations obtained in this book remain valid, provided the initial hypotheses made are verified (particularly that of a low volume fraction of particles). However, at high velocities or by the influence of heterogeneities in concentration, turbulence can occur, at the scale of the particle and/or that of the two-phase medium. Additionally, in the case of influence by significant external stresses such as variation of the flowrate, large-scale agglomerates may appear in the multiphase medium, as indeed may pockets of fluid. The agitation of the medium may become very great, and the equations established herein become insufficient to deal with the situation. Furthermore, this volume does not discuss transient phenomena; nor does it deal with cases where the size or concentration of particles are large, such as fluidized beds or granular media. Note that porous media are not discussed here either.

Hence, the ambition of this volume is very limited, but the few cases which are examined are fairly representative both of the complexity of the issue and of what is possible to achieve with relative ease.

Roger PRUD'HOMME
September 2014

List of Main Symbols

Latin characters

a, b partial derivatives of pressure with respect to ξ and $\mathbf{A}$ respectively

A chemical affinity; chemical species; monoatomic species

A_2 diatomic species

$\underline{\mathbf{A}}$ column matrix of chemical affinities in a multireactive medium

A, B Arrhenius coefficients

B Spalding parameter for mass transfer (B_M), or energy transfer (B_T), between a fluid mixture and an interface

c speed of sound; or speed of light in the void

$c_1, c_\mu, c_\varepsilon, c_k$ coefficients of the $k-\varepsilon$ method

C total number of moles per unit volume

C_j molar concentration per unit volume

C_p, C_v specific heat at constant pressure or at constant volume, respectively (c_p, c_v for the unit of mass)

d molecular droplet diameter; distance; or differential

D diffusion coefficient; or diameter

$D_{\alpha;i}$ diffusion coefficient of species α at quantic level i

$\vec{\vec{\mathbf{D}}}$ strain rate tensor

D_a Damköhler number

D_T thermal diffusion coefficient

$e_{v,\alpha}$ vibrational energy of diatomic species α

$\vec{\mathbf{e}}_i$ orthonormal basis vector

E internal energy (e per unit mass)

$E(k)$ energy spectrum of turbulence

E_a activation energy

f parameter; reduced chemical production rate; or Blasius function

f', f'' Reynolds, Favre fluctuation, respectively

$\boldsymbol{f}$ force acting on each unit mass

$\boldsymbol{f}_j$ force acting on the unitary mass of the species j

$f_\nu(r,s)$ volumetric distribution in particle radius, at abscissa s of a light beam of ν wave number

F Helmholtz free energy (f for the unit of mass); generalized force; or any extensive quantity (f for the unit of mass);

$\mathbf{F}$ force of F_x, F_y, F_z components

$\mathcal{F}$ source of impulsion due to molecular collisions

G Gibbs free enthalpy (g for the unit of mass)

$\mathbf{g}$ acceleration due to gravity (of modulus g)

g_j chemical potential per unit mass of the species j in a mixture

H enthalpy (h per unit mass)

$I_\nu(M,\mathbf{u})$ directional monochromatic radiative intensity at any M point in the $\mathbf{u}$ direction

$I_\nu^0(T)$ Planck function

$\mathbf{1}$ unit tensor

j,k chemical species

$\mathbf{J}$ flux with respect to an unmoving reference frame

$\mathcal{J}$ flux with respect to the barycentric motion of the fluid

$\mathcal{J}_{Dk}$ diffusion flux of the species k equal to $\rho_k(\mathbf{v}_k-\mathbf{v})$

k Boltzmann's constant; wave number; or kinetic energy $v^2/2$

$k(T)$ specific reaction rate

K kinetic energy (k per unit mass); number of chemical reactions in a mixture; heat exchange coefficient; or wave number

l length; latent heat per unit mass; or mean free path

ℓ transfer length; or integral scale of turbulence

ℓ_D diffusion thickness of a non premixed flame

ℓ_f, ℓ_δ respective thicknesses of preheating, of reaction of a premixed flame

ℓ_K length scales of Kolmogorov

L length; molar latent heat; number of chemical elements in a mixture; or phenomenological coefficient

Le Lewis number

L_{ij}, l_{ii} phenomenological coefficients

$\underline{\underline{\mathbf{L}}}$ matrix of phenomenological coefficients of chemical reactions

m total mass

M molecular mass; diluent; Mach number; or material point

$\mathcal{M}$ molar mass

m_j mass of the species j

$\mathcal{M}_j$ molar mass of the species j

$\dot{m}$ unit mass flow rate; mass flow rate; mass flow rate of a nozzle

n total number of moles

n_j number of moles of the species j

N number of species; number of molecules per unit volume; coordinate normal to an interface; or number of elements in a statistical calculus

$\mathcal{N}$ number of Avogadro

$\mathbf{n}, \mathbf{N}$ unitary normal to an interface; or to a surface

p thermodynamic pressure

P probability density

$\vec{\vec{\mathbf{P}}}$ pressure tensor

$P^R(M)$ radiative power at any point M, equal to $-\left(\nabla \cdot \mathbf{q}^R\right)(M)$

Pr Prandtl number

q parameter; or heat flux

$\dot{q}$ volume flow rate

$\mathbf{q}$ heat flux vector

$\mathbf{q}^R(M)$ radiative flux vector in M

q^{rad} radiative flux

Q partition function; quantity of heat; total enthalpy released per unit mass of pyrolysed fuel at the surface; or efficiency (Q_{abs} for absorption, Q_{sca} for scattering)

$\dot{Q}$ heat released per unit time at the walls of a chemical reactor; or lateral heat flux in a nozzle

$\left(Q_f^0\right)_j = \left(H_0^0\right)_j$ molar enthalpy of formation of the species j ($\left(q_f^0\right)_j$ per unit mass)

r perfect gas constant per unit mass; or radius

R molar universal gas constant; radius; or electrical resistance

$\mathfrak{R}$ reference frame

Re Reynolds number

S entropy (s per unit mass); area; or area of the cross section of a nozzle

(s) Arrhenius exponent

s curvilinear abscissa

s_L, s_L^0, s_t combustion velocity respectively laminar, standard, and turbulent

S, $\mathcal{S}$ surface

$\mathbf{S}$ symmetrical part of the velocity gradient tensor

Sc Schmidt number

$S_\nu(s', \mathbf{u})$ monochromatic source function

t time; or diffusion thickness

T absolute temperature

T_f flame temperature

T_a, T_{ad} activation temperature, adiabatic temperature of a reaction, respectively u, v, w velocity $\vec{\mathbf{v}}$ components in Cartesian coordinates (v_r, v_θ, v_z in cylindrical coordinates)

u unit direction vector

U, U_∞ reference velocity

U, **v** velocity vector; velocity vector (components v_x, v_y et v_z and modulus v) of a material point; or barycentric velocity vector in a composite fluid

v, v' velocity, turbulence intensity respectively

$\mathbf{v}_k$ velocity vector of species k

V velocity; force; or volume in the phase space

$\vec{\mathbf{V}}, \mathbf{V}$ vector; velocity vector; velocity vector in the phase space $(\vec{\mathbf{x}}, \vec{\zeta})$

$\mathcal{V}$ volume; control volume

$\mathbf{V}_k$ diffusion velocity vector of the species $\mathbf{v}_k - \mathbf{v}$

w velocity of a surface (normal component w); or $d\vec{\zeta}/dt$ in the phase space

W local velocity vector of a discontinuity

$\dot{W}_F$ rate of production of the quantity F

$\dot{W}_\alpha$ rate of production of species α

$\dot{W}_\alpha^d$ rate of production of species α due to the dispersed phase

$\dot{W}_{qdm}$ rate of production of impulsion

$\dot{W}_{E_t}$ rate of production of total energy

$\dot{W}_{rad}$ rate of production of total radiative energy

$\dot{W}_{E_t}^d$ rate of production of total energy due to the dispersed phase

$\dot{W}_{E\alpha,\text{int}}$ rate of production of energy for the internal degrees of freedom of the species α

$\dot{W}_{v,\alpha}$ rate of production of vibrational energy of the species α

$\dot{W}_{\alpha;i}$ rate of production of species α at quantic level i

$\dot{W}_{E_t a}$ rate of production of total energy per unit area and unit time at interface

x mass fraction of radiation absorbing species; or parameter of Mie

x, y, z Cartesian coordinates; x along a nozzle axis

$\vec{\mathbf{x}}$ position vector

X_j, Y_j molar and mass fraction of the species j, respectively

$Y_{\alpha;i}$ mass fraction of species α at quantic level i

Z fraction of mixture

Greek symbols

α species; α_ν directional monochromatic absorptivity

β_j, β_T reduced concentration, reduced temperature respectively

δ thickness of a viscous layer; $\delta(x)$ Dirac distribution

Δ difference; Laplacian; ΔH : heat of a reaction difference; ΔH_p : pyrolysis enthalpy per unit mass of pyrolyzed fuel

ε small dimensionless parameter; turbulent dissipation rate

ε_ν monochromatic emissivity

ϕ velocity potential; $\phi(\vec{\mathbf{x}}, t)$: weight function

φ scalar potential

$\Phi_\nu(s, \mathbf{u}' \to \mathbf{u})$ scattering phase function

$\phi^R(B)$ wall radiative flux (towards the fluid at a point B of the surface)

γ isentropic coefficient c_p/c_v ; or damping of a wave

η bulk viscosity; or reduced coordinate

$\eta_\nu(s)$ monochromatic emission coefficient at s

κ thermal diffusivity $\lambda/\rho c_p$; or mean curvature of a surface

$\kappa_\nu(s)$ monochromatic absorption coefficient at abscissa s of the light beam

λ coefficient of thermal conductivity; or eigenvalue

Λ coefficient of head loss; or heat transfer coefficient

μ coefficient of shear viscosity; Gibbs free energy per mole; or absorption coefficient per unit of wave length

μ_j molar chemical potential of a species j in a mixture

ν kinematic viscosity μ/ρ ; or light wavenumber

ν_j algebraic stoichiometric coefficient $\nu_j = \nu''_j - \nu'_j$

ν'_j, ν''_j stoichiometric coefficient of the direct reaction; or its inverse respectively

$\mathbf{\Pi}$ viscous pressure tensor

Π_i dimensionless ratio

θ temperature; or angular coordinate

ϑ volume per unit mass (inverse of the density)

ρ density (volumic mass)

ρ_j partial density of species j

$\rho_{\alpha;i}$ partial density of species α at quantic level i

σ surface tension

σ_ν monochromatic scattering coefficient

Σ surface; area of a surface; $\Sigma(x)$: area of the cross section of a nozzle

$\vec{\vec{\Sigma}}$ stress tensor

$\sigma_k, \sigma_\varepsilon$ Prandtl numbers of the $k-\varepsilon$ method

ϑ volume per unit mass

τ characteristic time; crossing time in a chemical reactor; dimensionless energy of reaction $\tau = \Delta H / c_p T_1$

$\tau_\nu(s',s)$ monochromatic transmissivity

ω speed of rotation; pulsation of an oscillating wave; or function of β_T

$\boldsymbol{\omega}, \boldsymbol{\Omega}$ rotation vector

Ω speed of rotation; or solid angle

ξ progress variable per unit mass; reduced coordinate; or correlation length

$\underline{\xi}$ column matrix of the progress variables

ψ stream function; or probability in the space phase

ζ stream function; probability in the space phase; or reduced variable

$\dot{\zeta}$ rate of production of a chemical reaction

$\boldsymbol{\zeta}$ vector of the phase space

Subscripts, superscripts, and other symbols

$_a$ of activation; relative to the quantities per unit area of the interface

$_{ad}$ adiabatic

b burnt gases

chem chemical

$CO{-}E_{\beta v}$ coupling CO molecule - vibrational energy

D direct; of dissociation; or diffusive

e equilibrium flow

eff effective

emi of emitted light

f frozen composition; fresh gases; flame

G , g gas

α, β, i, j of species

i internal; relative to imaginary part; irreversible

inci of incident light

int internal degrees of freedom of a molecule

l liquid

L line; liquid; laminar

leav relative to the light intensity leaving a wall

m mixture; or mass

mec mechanical

p at constant pressure; solid phase; or condensed particles

part particles

r chemical reaction; reference

R reverse; or recombination

s steady state; surface; isentropic; or specific

$_S$	surface; relative to the specific or intensive interfacial quantities
$_{\mathrm{st}}$	stoichiometric; or steady
$_t$	for translational energy mode of a molecule; or turbulent
$_T$	temperature; turbulent; or at constant temperature
^{T}	transposed of a tensor
$^\circ$	deviator of a tensor
$_{\mathrm{th}}$	thermal
$_u$	unburnt gases
$_v$, or $_\vartheta$	at constant volume
$_V$	vapor
$_{//}$	parallel to a surface
$_\perp$	normal to a surface
^{0}	standard reference value
$^\bullet$	pure simple substance
$\dot{}$	per unit time; or for a rate of production
$\bar{}$	average quantity; or Reynolds average
$\tilde{}$	transposed tensor; transposed matrix; or Favre average
$'$	Reynolds disturbance in relation to an average value
$''$	Favre disturbance in relation to an average value
$S(\ )$	symmetrical part of a matrix or a tensor
$\langle\ \rangle$	turbulent average
$(\ )_T^0$	standard thermodynamic function
$\times$	vector product

$\otimes$ tensor product

$\cdot$ scalar product (singly-contracted tensor product)

$:$ dyadic product (doubly-contracted tensor product)

$\wedge$ exterior product

$*$ sonic conditions; or reference state

$\vec{\nabla}$ nabla (gradient operator)

$[\ ]_-^+$ jump of a quantity across an interface

d/dt material derivative equal to $\partial/\partial t + \vec{\mathbf{v}} \cdot \vec{\nabla}$

$\partial/\partial t$ partial time derivative

$d_{\mathbf{W}}/dt$ material derivative associated with the velocity $\vec{\mathbf{W}}$ equal to $\partial/\partial t + \vec{\mathbf{W}} \cdot \vec{\nabla}$

$\rightleftarrows$ for a reversible chemical reaction

1

Generation of Multiphase Flows

Multiphase mixtures exist in the natural state, and are highly varied. If we wish to generate multiphase flows for research or for industrial purposes, it is not always easy to do, and often necessitates a profound technological study. In this chapter, we examine the issue of the generation of particular two-phase mixtures: dusty gases, fogs and bubbling flows. The production of mists for combustion is a complex problem, with numerous steps, usually divided into primary atomization and secondary atomization.

1.1. Creation of suspensions of solid particles in a gaseous phase

Here we present two very different examples of the generation of flows with solid particles.

The first case study is designed as a micro-gravity examination of the combustion of a suspension of wheat starch particles.

The second study concerns the formation of soot.

1.1.1. *Creation of a homogeneous suspension of starch particles*

Let us look at the example of the creation, for research purposes[1], of a suspension of wheat starch particles of 20μm average diameter in a vessel [VEY 03, BOZ 04, TAK 09, BOZ 10].

Consider a specific setup of apparatus, comprising a transparent cylindrical chamber with an octagonal cross-section and short relative length (volume = 20 liters; length/transverse dimension ≡ 2.2). The particles are dispersed by means of a turbulent flow created by the discharging of an auxiliary tank (see Figure 1.1). The experiments are conducted with starch/air mixtures at concentrations lower than $400 g/m^3$.

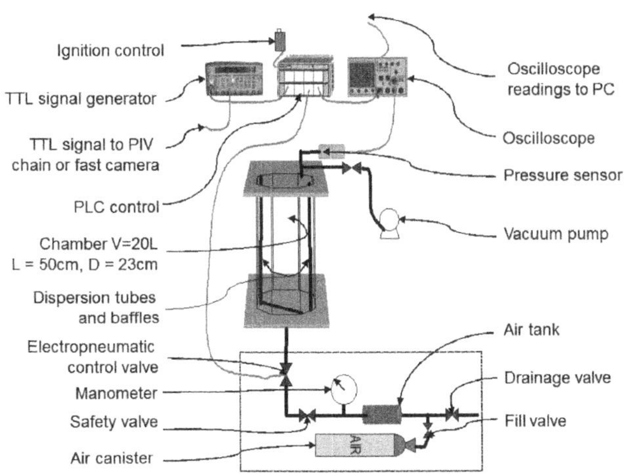

Figure 1.1. *Experimental setup for the generation of a suspension of solid particles [VEY 03] (diagram reproduced with kind permission from the authors)*

1 The objective is to study the mechanisms of combustion in suspensions of fine solid particles, dispersed in a gaseous medium, by attempting to correlate the characteristics of isochoric explosion with the initial conditions existing in the mixture at the moment of ignition.

The recordings, made by fast imaging techniques (LDV and PIV[2]), of the evolution over time of the aerodynamic state of the mixture and the distribution of particles in the chamber indicate that the time needed to obtain an optimal mixture (minimal velocity and turbulence intensity, quasi-homogeneous concentration) is between 500 and 700 ms.

1.1.2. *Soot formation*

Soot particles are formed during the combustion of carbonate species. Their nucleation and growth, and their interaction with the flow of reagents, have been studied by many authors. In particular, we can cite the case of the Emmons problem, involving a boundary layer with a flame above a flat plate [EMM 56, WIL 85] (see section 5.3.2). The relatively recent effort to take account of soot particles in this problem has required appropriate mechanisms to be developed [BOC 94, FRE 89, FRE 94, WAN 96, ZHA 05, SGR 07], and it has been necessary to take into account the soot's interaction with the flow and the ablative surface [WAN 07, LEG 05b, LEG 06, LEG 09].

Dorey [DOR 12] offers a complete review of this complex problem. There are four distinct processes involved:

1) nucleation;

2) surface growth;

3) oxidation;

4) coagulation.

Let us now summarize these steps as simply as possible.

Nucleation occurs because of the degradation of gaseous hydrocarbons by pyrolysis, which leads to the formation of smaller carbonate species, such as acetylene C_2H_2 and

2 LDV: *Laser Doppler Velocimetry*; PIV: *Particle Image Velocimetry*.

benzene. A wide variety of large, flat molecules known as PAHs (*Polycyclic Aromatic Hydrocarbons*) can be created by the coming together of these species.

The formulation proposed by Frenklach and Wang [FRE 94] for the nucleation of PAHs is given by the system of chemical reactions [1.1]:

$$\begin{array}{lcl} A_i + H & \rightleftarrows & A_i^{\cdot} + H_2 \\ A_i^{\cdot} + C_2H_2 & \rightleftarrows & A_iC_2H_2^{\cdot} \\ A_iC_2H_2^{\cdot} + C_2H_2 & \longrightarrow & A_{i+1} + H \end{array} \qquad [1.1]$$

In this system, the species A_i represents a molecule containing i aromatic rings fused together. $A_i^{\cdot}$ is the associated radical and $A_iC_2H_2^{\cdot}$ is the radical obtained after the addition of acetylene.

For the *growth of surface area* of soot, Frenklach and Wang [FRE 94] present the analogy with the HACA mechanism (addition of acetylene by the reaction: "H-Abstraction/C_2H_2-Addition") for the surface of the PAHs, and obtain the system [1.2], where C_S represents a soot particle and $C_S^{\cdot}$ the radical associated therewith:

$$\begin{array}{lcl} C_S + H & \rightleftarrows & C_S^{\cdot} + H_2 \\ C_S^{\cdot} + H & \longrightarrow & C_S \\ C_S^{\cdot} + C_2H_2 & \longrightarrow & C_S + H \\ C_S^{\cdot} + C_2H_2 & \longrightarrow & C_S^{\cdot} + H_2 \end{array} \qquad [1.2]$$

Oxidation is characterized by the two reactions:

$$\begin{array}{l} C_S^{\cdot} + O_2 \rightarrow \text{products} \\ C_S + OH^{\cdot} \rightarrow \text{products} \end{array} \qquad [1.3]$$

The species O, CO_2 and H_2O may also be involved, but play a lesser role in the oxidation process [XUE 03].

Coagulation results from collisional interactions between the small particles thus formed. The process involves coalescence (fusion of two particles into one) and agglomeration (particles which stick to one another).

1.2. Creation of suspensions of bubbles in a liquid

Bubbly flows have been studied for a long time, hence we now have a deep understanding of their behavior, particularly in pipes [CLI 78, NIG 91] and around submarine propellers.

The nucleation of bubbles of vapor can hinder heat transfer when pipes are cooled. This is a major problem for the cooling circuits of rocket engines. For example, the walls of the nozzles of cryogenic rocket engines are lined with tubes filled with liquid hydrogen, and the appearance of bubbles of gaseous hydrogen can block the cooling process and lead to the destruction of the nozzle.

Bubbly flows are very sensitive to gravity.

The gas bubbles may also be of different substances than the liquid, as is the case with air bubbles in water. However, it is also possible for the foreign gas to be partially soluble in the liquid, such as carbonic gas in water.

1.2.1. *Example of creation of a suspension of bubbles in a liquid*

Let us now discuss the example of the generation of a two-phase water/air flow for the purposes of a scientific experiment. This is the "effervescent atomization" examined by Lhuissier [LHU 11] as a means of obtaining liquid filaments.[3] To obtain a satisfactory suspension of bubbles, Lhuissier uses an experimental device with an overflow tank: the water falls onto a turbine, below which is a porous wall through which air is injected. The millimetric air bubbles obtained are broken up into smaller bubbles by the rapid rotation of the turbine blades and the counter-blades fixed to the walls. The suspension is channeled into a sorting chamber where the largest bubbles are eliminated by gravity (this mechanism exploits the difference in rising speed, of the order $\rho g d_b/\mu$, where d_b is the diameter of the bubbles and μ is the viscosity of the solution, containing a surfactant to prevent coalescence). Then the suspension flows into an injection nozzle which feeds the actual experiment (see Figure 1.2).

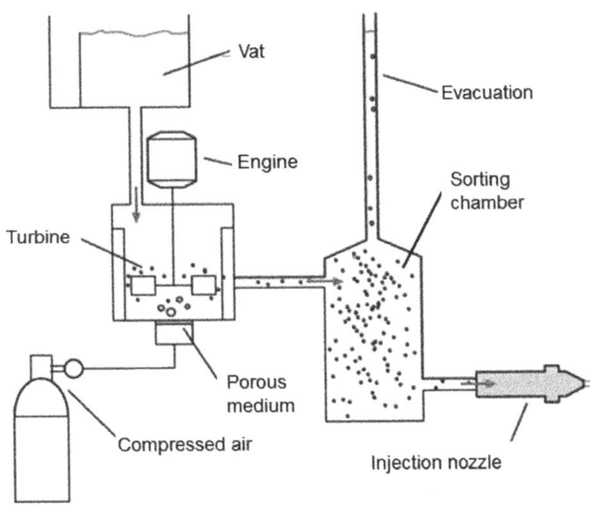

Figure 1.2. *Generation of a suspension of bubbles (extracted from a diagram in Lhuissier [LHU 11], with the author's permission)*

3 The formation of liquid filaments and their subsequent disaggregation into droplets is an important step in the process of atomization of concentric liquid jets.

1.2.2. *Influence of gravity on suspensions in pipes*

Bubbly flows are a particular case of two-phase gas/liquid flows. As one might expect, the influence of gravity is important, as shown in Figure 1.3 [LEG 99].

If these flows take place in pipes, the gaseous phase may arrange itself as a suspension of small, almost-spherical bubbles under the influence of the surface tension. However, it may also be arranged differently. Thus, for significant volume fractions of gas, it is possible to obtain "*slug flows*", with large bubbles greater in size than the diameter of the pipe itself, or sometimes even gaseous flows surrounded by liquid films, moving along the pipe.

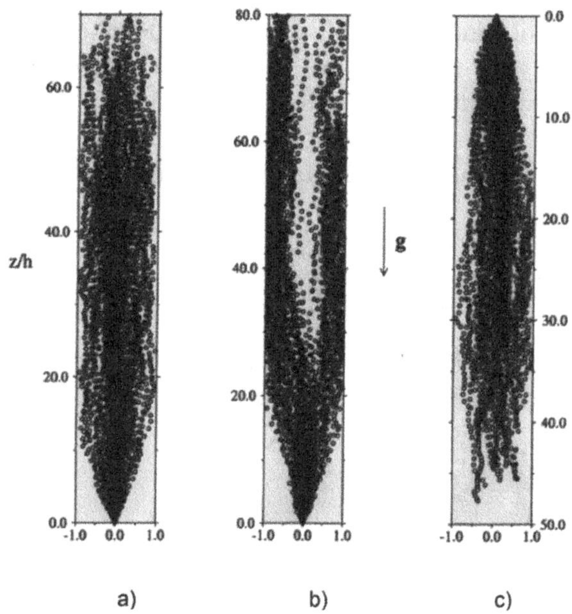

Figure 1.3. *Trajectories calculated by DNS for bubbles of 0.5mm in diameter, in a turbulent flow of Re = 40,000, in a pipe of length 2m and half-thickness 2cm: a) microgravity conditions; b) normal terrestrial gravity, ascending vertical flow; c) normal terrestrial gravity, descending vertical flow. The DNS results conform closely to the experimental results found by [LEG 99] (diagram reproduced with the kind permission of the authors)*

1.2.3. *Slug flows*

Slug flows are intermittent flows. In the context of the *UC concept* (UC standing for “unit cell”) [NIC 62, WAL 69], the unitary part of the *slug flow*, in the inclined pipe in Figure 1.4(c) for example, is a cell containing a long bubble (a “pocket”) of length L_S, and a liquid “slug” of length L_D, including dispersed bubbles. On average the flow is quasi-steady in a frame of reference which shifts at average velocity V of large bubbles.

During a stable *slug flow*, the liquid feeds into the slug from behind at the same rate as it exits at the front. Hence, the length of the slug remains constant as it moves along the pipe. For quasi-horizontal pipes (see Figure 1.4(b)), the ingoing liquid from behind the slug slows down under the influence of wall friction and forms a stratified layer. In the case of quasi-vertical pipes (see Figure 1.4(a)), a ring-shaped film descends, increasing in speed as it does so. The majority of the gas is transported in the large bubbles (or pockets).

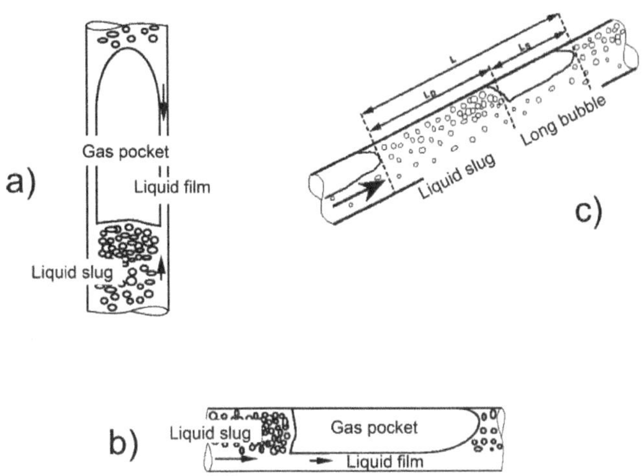

Figure 1.4. *Pipe flow, respectively: a) vertical, b) horizontal, c) inclined, showing the existence of zones with pockets of gas and liquid films, alternating with slugs in bubbly suspensions (figure provided by J. Fabre; also see Fabre and Liné [FAB 92])*

A 1D simulation has been drawn up, with a steady-state system of differential equations, their boundary conditions and constitutive relations [FAB 92].

The results have successfully been compared to the experimental observations, thus validating the *UC concept*.

1.3. Creation of suspensions of drops in a gas

The problems of atomization are crucially important for the combustion of liquid propellants. It is for this reason that particular attention is paid to this issue in this book on flows with chemical reactions.

Here, we shall examine sprays and the atomization of liquid fuels upon injection into internal combustion engines [EGG 97, FER 09].

The linearized study of semi-infinite layers of perfect fluids offers us the opportunity to examine the Rayleigh–Taylor and Kelvin–Helmholtz instabilities. The case of an initially-flat thin layer of liquid which is destabilized by the influence of an external flow parallel to the plane of that liquid layer is then presented. Next we look at thin layers and the influence of turbulence. These instabilities lead to the formation of liquid filaments – an intermediary step before the formation of drops, which we will go on to examine.

In certain cases, primary atomization may be followed by secondary atomization, during which the primary drops divide and flocculate to reach their local distribution in a state of macroscopic suspension.

1.3.1. *Destabilization of fluid sheets and layers*

1.3.1.1. *Linear study of the instabilities of semi-infinite layers of perfect fluids*

Consider superposed flows of incompressible fluids [LAN 71], with velocity potentials such that:

$$\nabla \times \mathbf{v} = \mathbf{0}, \quad \mathbf{v} = \nabla \phi, \quad \nabla \cdot \mathbf{v} = 0 \rightarrow \nabla^2 \phi = 0 \qquad [1.4]$$

In the reference state, the separating surface is flat and horizontal ($y = 0$), and fluid (2) is laid on top of fluid (1). We wish to study the motion of small disturbances in the two-dimensional space, such that:

$$\phi = \phi^* + \phi', \quad \phi' = \hat{\phi} e^{i(kx + hy - \omega t)} \qquad [1.5]$$

where the asterisk denotes the reference value (Figure 1.5) and where h and k are wavenumbers and ω the pulsation of the wave. *A priori*, we assume that k is real and positive, and that h and ω are complex numbers. The velocity vector $\mathbf{v}$ has the components (u, v).

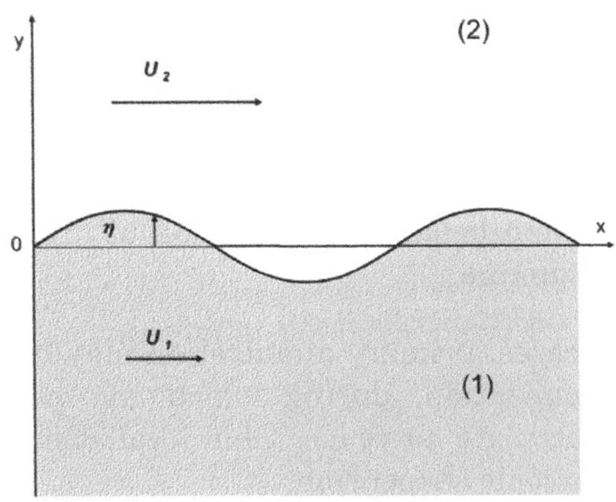

Figure 1.5. *Disturbance waves of a surface separating two superposed fluid layers (1) and (2), of semi-infinite thicknesses, with reference velocities:* $\mathbf{v}_1 = (U_1, 0)$, $\mathbf{v}_2 = (U_2, 0)$. *The elevation of the disturbed interface is* $\eta(x, t)$

Equations [1.4] and [1.5] give us:

$$\begin{cases} h^2 + k^2 = 0, \\ \phi_1 = U_1 x + \phi'_1 = U_1 x + \alpha_1 e^{ky} e^{i(kx-\omega t)} \\ \phi_2 = U_2 x + \phi'_2 = U_2 x + \beta_2 e^{-ky} e^{i(kx-\omega t)} \end{cases} \quad [1.6]$$

The constants α_1 and β_2 are determined using the equations at the disturbed interface (e.g. see [DUD 76]) where, assuming that interface is impermeable and the Laplace law is valid, we have:

$$\begin{cases} y = \eta(x,t) = a e^{i(kx-\omega t)}, \quad p'_1 - p'_2 = -\sigma \eta_{,xx} \\ v_1 - U_1 \eta_{,x} = \eta_{,t}, \quad v_2 - U_2 \eta_{,x} = \eta_{,t} \end{cases} \quad [1.7]$$

where $y = \eta(x,t)$ is the equation of the disturbed interface $y = 0$ in the reference state, σ is the surface tension, and p'_1, p'_2 are the local disturbances in pressure.

Using the Bernoulli equations, we also find, at the interface:

$$\begin{cases} \phi'_{1,t} + U_1 \phi'_{1,x} + g\eta + p'_1/\rho_1 = 0 \\ \phi'_{2,t} + U_2 \phi'_{2,x} + g\eta + p'_2/\rho_2 = 0 \end{cases} \quad [1.8]$$

where g is the acceleration due to gravity and ρ_1, ρ_2 are the volumetric densities of the fluids, which can be taken to be constant.

If we eliminate the pressure terms, we find the following interface conditions:

$$\begin{cases} \rho_2 \left(\phi'_{2,t} + U_2 \phi'_{2,x}\right) - \rho_1 \left(\phi'_{1,t} + U_1 \phi'_{1,x}\right) + \left(\rho_2 - \rho_1\right) g\eta + \sigma \eta_{,xx} = 0 \\ \phi'_{1,y} - U_1 \eta_{,x} = \eta_{,t}, \ \phi'_{2,y} - U_2 \eta_{,x} = \eta_{,t} \end{cases} \quad [1.9]$$

The dispersion equation can easily be deduced from this:

$$\left(\omega - k\frac{\rho_1 U_1 + \rho_2 U_2}{\rho_1 + \rho_2}\right)^2 = \frac{\sigma}{\rho_1 + \rho_2}k^3 - \frac{\rho_1 \rho_2 (U_2 - U_1)^2}{(\rho_1 + \rho_2)^2}k^2 - g\frac{\rho_2 - \rho_1}{\rho_1 + \rho_2}k \qquad [1.10]$$

We then separate the real and imaginary parts, and because the wavenumber is assumed to be real and positive, we obtain:

$$\begin{cases} \omega_r = k\dfrac{\rho_1 U_1 + \rho_2 U_2}{\rho_1 + \rho_2} \\ \omega_i^2 = \dfrac{\sigma}{\rho_1 + \rho_2}k^3 - \dfrac{\rho_1 \rho_2 (U_2 - U_1)^2}{(\rho_1 + \rho_2)^2}k^2 - g\dfrac{\rho_2 - \rho_1}{\rho_1 + \rho_2}k \end{cases} \qquad [1.11]$$

This mirrors the classic results relating to instabilities (between perfect fluids) found by *Rayleigh–Taylor* (if we set: $U_1 = U_2$) and *Kelvin–Helmholtz* (if we suppose that $g = 0$). It is possible to study the influence of the surface tension [GAT 01].

1.3.1.2. *Linear study of the instabilities of a layer of viscous liquid, of finite thickness, in the presence of flows of incompressible perfect gases*

In order to examine the destabilization of a flat sheet before breakup, we can conduct a linearized study [BRE 07]. Here, we consider the case of a layer of viscous liquid (1) of height h, flat in the reference state, in the presence of a gaseous flow (2), parallel to the plane, with the same velocity on both sides of the liquid (see Figure 1.6).

The small disturbances of the flow parameters are of the form:

$$f' = \hat{f}\, e^{i(kx + ly - \omega t)}$$

(where $l = m$ for the stream function) and the disturbed liquid–gas interfaces are defined by:

$$y = \eta(x,t) \pm \frac{h}{2}$$

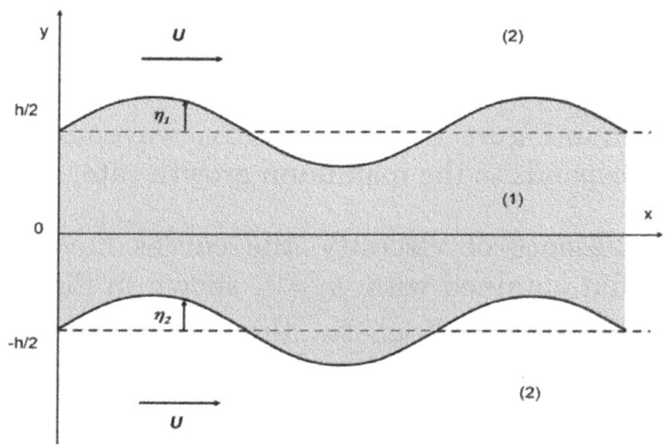

Figure 1.6. *Sinusoidal waves of a liquid layer, which is flat in the reference state*

We find the following dispersion equation linking the wavenumber k to the pulsation ω [COU 96]:

$$a_1 (\Omega+2)^2 - 4\sqrt{\Omega+1}\,[\coth(mh) - \frac{1}{sh(mh)}] + \rho\,\Omega^2 + 2i\rho\,\mathrm{Re}\,\Omega + a_2 = 0 \quad [1.12]$$

where the following reduced quantities play a role:

$$\Omega = \frac{i(kU_1 - \omega)\rho_1}{\mu_{1k}}, \quad \mathrm{Re} = \frac{\rho_1 (U_2 - U_1)}{\mu_1 k}$$

$$\rho = \frac{\rho_2}{\rho_1},\ a_1 = \coth(kh) - \frac{1}{sh(kh)} = th\left(\frac{kh}{2}\right),\ a_2 = \frac{\rho_1 \sigma}{\mu_1 k} - \rho\,\mathrm{Re}^2$$

and, in particular, the Reynolds number Re and Ohnesorge number Oh:

$$Oh = \frac{\rho_1 \sigma}{\mu_1 k}$$

Given that $\omega = \omega_r + i\,\omega_i$, we can then trace the curves giving the growth rate ω_i as a function of k (an example of this is given in Figure 1.7). The effective wavenumber is that which corresponds to the maximum growth rate.

In the absence of viscosity, the curves have the same shape as that obtained with $\mu_L = 0$, shown in Figure 1.7, for $\sigma = 0,075 N/m$. According to the study conducted by Cousin and Dumouchel [COU 96], when the viscosity and surface tension are both taken into account, the surface tension controls the order of growth of the disturbances (when the Weber number $We = \rho_L U^2 h/2\sigma$ is less than 1, no sinusoidal wave can grow at the liquid interfaces so long as $\rho = \rho_2/\rho_1 < 1$), whilst the viscosity imposes the characteristics of the dominant wave. We can also see that the inflection points do not feature on the dispersion curves, at a given level of surface tension, unless the viscosity value is sufficiently high.

The linearized method can be applied to many other configurations of liquid layers in the presence of gas flow (in particular, see [LAM 45, COU 96a, PRU 98, GAT 01]).

The breakup of a liquid sheet by sinuous waves or dilatation waves was examined by Dombrowski and Hooper [DOM 62], and in particular the breakup of a liquid filament in Rayleigh mode is described by Raynal [RAY 97].

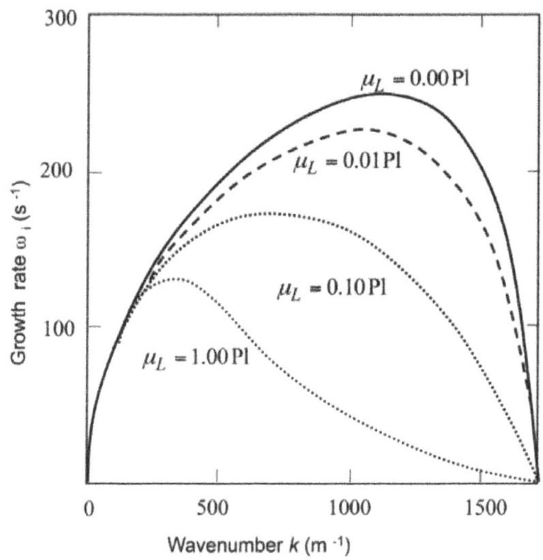

Figure 1.7. *Diagrams of dispersion as a function of the viscosity in Poiseuille mode in the case: h = 3mm, U = 10ms*$^{-1}$*, ρ = 1.29 × 10*$^{-3}$*kg.m*$^{-3}$*, σ = 0.075Nm*$^{-1}$ *(reproduced with kind permission from C. Dumouchel – also see [COU 96b])*

1.3.1.3. *Linear study of the instabilities of a thin film of viscous liquid in the presence of a flow of incompressible perfect gas*

We shall consider a liquid film resting on a flat plate, as done by Oron [ORO 97], and later Lalo [LAL 06].

Oron gives the equation which governs the motion of this film of height h (reference height h_0) subject to small sinusoidal disturbances (see Figure 1.8):

$$\mu\frac{\partial h}{\partial t}+\frac{\partial}{\partial x}\left[\left(\tau_{//}+\frac{\partial\sigma}{\partial x}\right)\left(\frac{1}{2}h^2+\beta h\right)-\left(\frac{1}{3}h^3+\beta h^2\right)\frac{\partial\phi}{\partial z}\bigg|_{z=h}-\sigma\frac{\partial^2 h}{\partial x^2}-\tau_{\perp}\right]=0 \quad [1.13]$$

where β is the slip coefficient of the liquid phase (zero for a continuous film), ϕ is the potential of the conservative force (gravity), $\tau_{//}, \tau_{\perp}$ the tangential stress and normal stress imposed at the interface, and h the thickness (or height) of the film. The analysis is performed by searching for solutions for the disturbance of thickness written $h' = h_0 e^{ikx+st}$ where k and s are real.

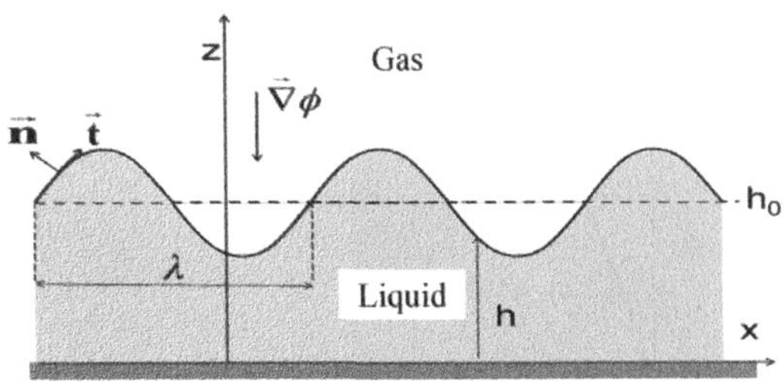

Figure 1.8. *The liquid film in Oron's study (redrawn from [ORO 97])*

Equation [1.13] is simplified in the case where $\beta = \tau_\perp = \phi = \partial\sigma/\partial x = 0$ and the shear force on the gas is constant:

$$\mu\frac{\partial h}{\partial t} + \tau_{//}\, h\frac{\partial h}{\partial x} + \frac{1}{3}\sigma h^3 \frac{\partial}{\partial x}\left(h^3 \frac{\partial^3 h}{\partial x^3}\right) = 0 \qquad [1.14]$$

In the absence of shear, but in the presence of gravity:

$$\beta = \tau_\perp = \tau_{//} = \partial\sigma/\partial x = 0, \quad \phi = \rho g z$$

we have:

$$\mu\frac{\partial h}{\partial t} - \frac{1}{3}\rho g \frac{\partial}{\partial x}\left(h^3 \frac{\partial h}{\partial x}\right) + \frac{1}{3}\sigma h^3 \frac{\partial}{\partial x}\left(h^3 \frac{\partial^3 h}{\partial x^3}\right) = 0 \qquad [1.15]$$

We find the same results as mentioned previously in section 1.3.1.1 for the Rayleigh–Taylor instability, this time in the presence of viscosity.

If gravity is orientated away from the surrounding mixture towards the liquid, there is no instability. If it is orientated the other way, however, the film's behavior is unstable for $0 < k < k_c$ where $k_c = \rho|g|h_0/\sigma = Bo/h_0$ is the critical wave number beyond which the system becomes stable and Bo is the Bond number $Bo = \rho|g|h_0^2/\sigma$.

1.3.1.4. *Droplet generation by vibrations in a direction normal to the liquid layer*

We propose to generate droplets by vibration of a recipient in the direction normal to the liquid layer (as illustrated by Figure 1.9). The resulting instability is called parametric instability or *Faraday instability* [LAN 62, BOU 09, PAR 08].

The process described below is that devised by Benjamin and Ursell [BEN 54], and then repeated by Marie Lalo [LAL 06].

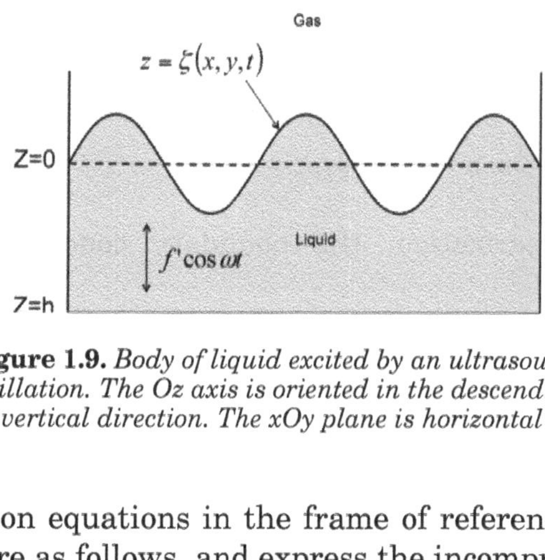

Figure 1.9. *Body of liquid excited by an ultrasound oscillation. The Oz axis is oriented in the descending vertical direction. The xOy plane is horizontal*

The motion equations in the frame of reference linked to the vessel are as follows, and express the incompressibility of fluids supposed to be non-viscous:

$$\begin{cases} \nabla \cdot \mathbf{V} = 0, \ \mathbf{F} = \rho\left(g - f'\cos\omega t\right) \mathbf{e}_z \\ \dfrac{\partial \mathbf{V}}{\partial t} + \mathbf{V} \cdot \nabla \otimes \mathbf{V} + \dfrac{1}{\rho}\nabla p + \dfrac{1}{\rho}\mathbf{F} = \mathbf{0} \end{cases} \quad [1.16]$$

The hypothesis of a velocity potential: $\mathbf{V} = \nabla\phi$ enables us to deduce, from the system [1.16]:

$$\begin{cases} \nabla^2\phi = 0, \\ \dfrac{\partial\phi}{\partial t} + \dfrac{|\mathbf{V}|^2}{2} + \dfrac{p}{\rho} - (g - f'\cos\omega t)z = \varphi(t) \end{cases} \quad [1.17]$$

The boundary conditions need to be written for the free surface area, on the side walls and at the bottom:

$$\begin{cases} \text{- free surface area:} \\ z = \zeta(x,y,t) \\ \dfrac{\partial\zeta}{\partial t} + u\dfrac{\partial\zeta}{\partial x} + v\dfrac{\partial\zeta}{\partial y} + w = 0 \ (\textit{kinematic condition}) \\ \dfrac{\partial\phi}{\partial t} + \dfrac{|\mathbf{V}|^2}{2} + \dfrac{\sigma}{\rho}\left(\dfrac{\partial^2\zeta}{\partial x^2} + \dfrac{\partial^2\zeta}{\partial y^2}\right) - (g - f'\cos\omega t)\zeta = 0 \\ \text{- side walls: } u = v = 0 \text{ or: } \partial\phi/\partial n = 0 \\ \text{- bottom: } w = 0 \text{ or: } \partial\phi/\partial z = 0 \end{cases} \quad [1.18]$$

After linearization, the boundary conditions at $z = 0$ become:

$$\begin{cases} \left.\dfrac{\partial\phi}{\partial t}\right|_{z=0} + \dfrac{\sigma}{\rho}\left(\dfrac{\partial^2\zeta}{\partial x^2} + \dfrac{\partial^2\zeta}{\partial y^2}\right) - (g - f'\cos\omega t)z = 0 \\ \dfrac{\partial\zeta}{\partial t} = \dfrac{\partial\phi}{\partial z} \end{cases} \quad [1.19]$$

Let us use the notation $k_m^{\ 2}$ to denote the eigenvalues of the Laplacian operator of surface $\left(\vec{\nabla}_{//}\right)^2 = \dfrac{\partial^2}{\partial x^2} + \dfrac{\partial^2}{\partial y^2}$ and $S_m(x,y)$ for the independent eigenfunctions such that:

$\left(\frac{\partial^2}{\partial x^2}+\frac{\partial^2}{\partial y^2}+k_m{}^2\right)S_m(x,y)=0$. We decompose the solutions of the system [1.19] on the basis of the $S_m(x,y)$:

$$\zeta(x,y,t)=\sum_0^\infty a_m(t)S_m(x,y) \qquad [1.20]$$

The independence of eigenfunctions means that we can obtain the following (see [LAL 06] for details of the calculations):

$$\frac{d^2a_m}{dt^2}+k_m\text{th}(k_mh)\left(\frac{k_m{}^2\sigma}{\rho}+g-f'\cos\omega t\right)a_m=0 \qquad [1.21]$$

Thus, by defining:

$$p_{m=}\frac{4k_m\text{th}(k_mh)}{\omega^2}\left(\frac{k_m{}^2\sigma}{\rho}+g\right),\ q_m=\frac{2k_mf\text{th}(k_mh)}{\omega^2} \qquad [1.22]$$

We have $f'=\alpha\omega^2$, the amplitude of oscillation is q_m and $p_m=4\omega_m{}^2/\omega^2$, where ω_m is the pulsation of the natural mode satisfying the dispersion equation:

$$\omega_{m=}\sqrt{k_m\text{th}(k_mh)\left(\frac{k_m{}^2\sigma}{\rho}+g\right)} \qquad [1.23]$$

By setting $T=\omega t/2$, equation [1.21] becomes the Mathieu–Hill equation:

$$\frac{d^2a_m}{dT^2}+(p_m-2q_m\cos 2T)a_m=0 \qquad [1.24]$$

whose solution is of the form [SIN 95]:

$$a_m(T) = e^{i\mu T}\Theta(T),\ \mu = \mu(p_m, q_m) \qquad [1.25]$$

where Θ is a periodic function.

The stability diagram put forward by Benjamin and Ursell [BEN 54] is given in Figure 1.10 in the plane (p_m, q_m). It includes several separate zones of instability. The stability of the free surface of a liquid subject to a forced oscillation depends on the excitation imposed (acceleration f' and pulsation ω) and on the properties of that liquid (surface tension σ and density ρ).

Figure 1.11 is deduced from a photograph of atomization by vibration normal to the liquid layer.

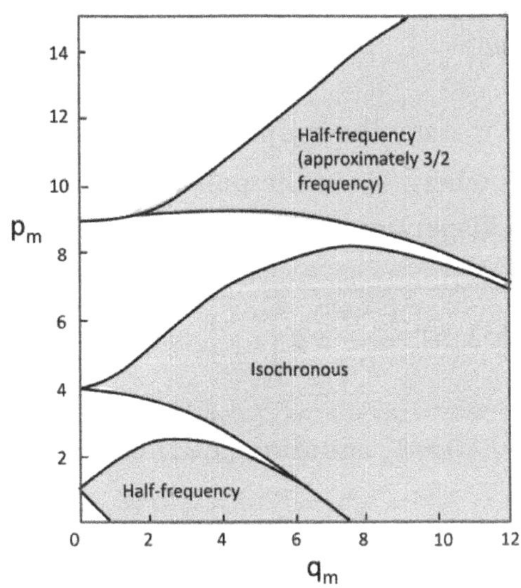

Figure 1.10. *Diagram of stability of a liquid layer subjected to forced oscillation (figure redrawn from [BEN 54])*

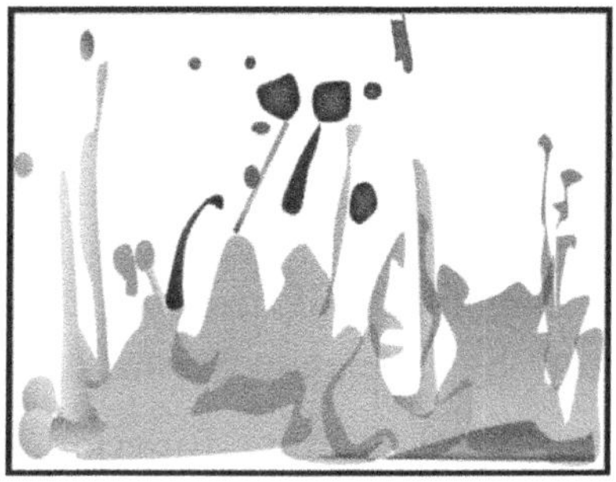

Figure 1.11. *Diagram of the atomization of a distilled water solution with 80% glycerine ($\nu = 0.43 cm^2 / s$) at the frequency of 20Hz. Figure drawn from a photo in Goodridge [GOO 97], showing a highly disturbed interface with numerous filaments, some of which form droplets*

1.3.1.5. *Generation of filaments from a liquid sheet seeded with air bubbles*

Using the experiment mentioned in section 1.2.1, it is possible to obtain a sheet seeded with air bubbles, as seen in Figure 1.12, which is the continuation of Figure 1.2, drawn from Lhuissier [LHU 11].

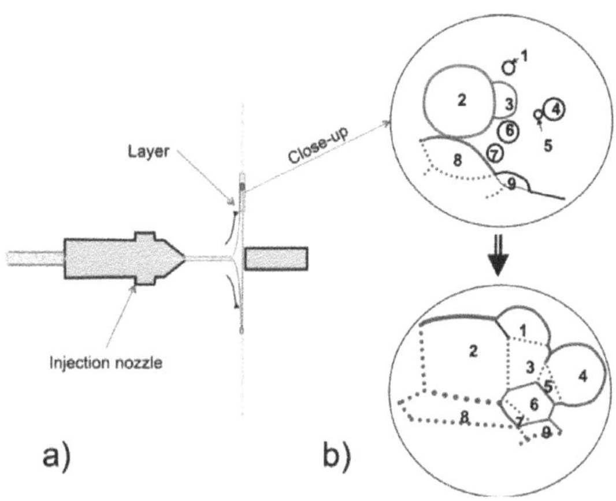

Figure 1.12. *Diagram showing effervescent atomization: a) the injection nozzle and the bubbly sheet of liquid formed along a plane at rest; b) formation, in that sheet, of holes bordered by filaments of increasing thickness (solid lines) which fragment into strings of droplets (dotted lines) at two successive times (redrawn from [LHU 11])*

The sheet which forms causes holes to appear, which grow with beading around their edges (accumulating the liquid thus liberated). Adjacent beads fuse into a network of filaments, from which drops are formed.

The intrinsic dispersion of the breakup of a filament (called a "ligament" by Lhuissier) around the mean $\langle d \rangle_{lig}$ is closely represented by a gamma function, the nth order of which determines the breadth:

$$S_n\left(x = \frac{d}{\langle d \rangle_{lig}}\right) = \frac{n^n}{\Gamma_f(n)} x^{n-1} e^{-nx} \qquad [1.26]$$

With a low value of *n* (i.e. a broad distribution), for an isolated filament, we have:

$$r(d) = \int_0^\infty \tilde{S}_n\left(\frac{d}{b_n d_l}\right) q(d_l) \frac{1}{b_n d_l} d(d_l) \qquad [1.27]$$

where $b_n = d / \langle d \rangle_{lig}$.

With a high value of *n* and therefore a smooth filament, the breakup is monodispersed and:

$$\begin{cases} \tilde{r}_0\left(x = \dfrac{d}{\langle d \rangle}\right) = \dfrac{4x^2}{\left[\Gamma_f(3/4)\right]^4} \exp - \left[x / \Gamma_f(3/4)\right]^4 \\ \langle d \rangle_0 = \dfrac{2\sqrt{2}}{\pi^{3/4}\, \Gamma_f(3/4) b_0} \sqrt{\langle l \rangle h} \end{cases} \qquad [1.28]$$

1.3.2. *Formation of droplets from filaments*

As we have just seen in Figure 1.11, regarding parametric instability, in general, drops do not form directly from the liquid body. Instead, that liquid body first forms filaments.

Drops are then formed from these liquid filaments, as we can see in Figure 1.12, which shows effervescent atomization, and Figure 1.13, where we show what happens in the presence of shear instability.

Note that if a liquid cylinder becomes detached (see Figures 1.13(a) and 1.13(b)) and is transported by the fluid at the speed of the wave, then the drop formation will obey a Rayleigh process[4] [DOM 54, DOM 62, RAY 78, TIO 93, OLE 97]. If, on the other hand, the liquid cylinder remains attached into two branches [RAY 97], then its atomization will result from the external flow, whose velocity is similar to that of the gas [DOM 63, KRÜ 90, SIT 85].

When the filaments break in Rayleigh mode, the pressure field, induced by capillary effects, induces a fluid motion. Any small disturbance causes variations in the straight section of the cylindrical filament and in the resulting flow which results in the accumulation of liquid in the regions with a larger straight cross-section, until drops form. The wavelength deduced from the Rayleigh analysis (see section 1.3.2.1) is equal to 4.51 times the diameter of the filament, and the law of conservation of volume means that the diameter of the drops (supposed to be spherical) is equal to 1.89 times the diameter of the filament.

In the case of Figure 1.13(c), the filaments remaining attached at the crest of the wave align in the direction of the flow whilst retaining their initial velocity. We can give an estimation of the jump in velocity between the filament

4 In the Rayleigh rupture regime, the aerodynamic effects are negligible. The rupture of the filaments is caused by the effect of surface tension, which acts to minimize the surface free energy of the liquid/gas interface by forming drops. Rayleigh [RAY 78] was the first to model the phenomenon in terms of small disturbances in 1878.

branch and the external gas flow [DIM 86]. The application of the small perturbations method thus yields the characteristic wavelength [KRÜ 90], and a drop diameter supposed to be proportional to that wavelength. From this an average diameter can be deduced.

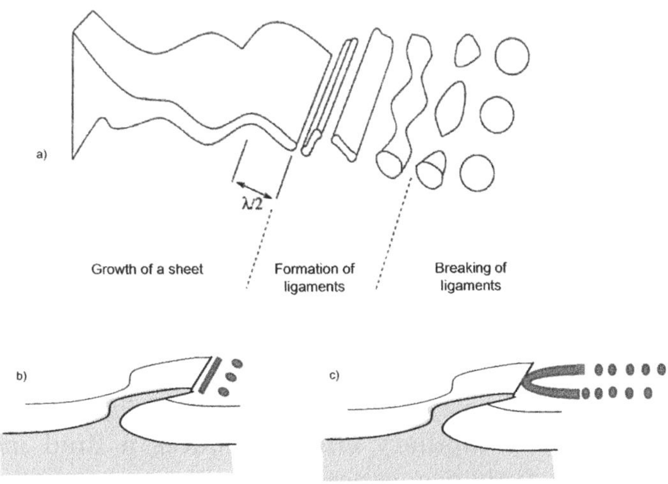

Figure 1.13. *Successive mechanisms leading to drop formation: a) liquid film; b) thin layer: detached transverse cylinder; c) thin layer: attached pair of filaments*

1.3.2.1. *Linear study of the instability of an isolated perfect liquid cylinder*

Exactly as we did with the liquid layers discussed in section 1.3.1, we shall begin, here, by examining ideal cases. The liquid cylinder considered here, created by a cylindrical pipe, is semi-infinite. Locally, it satisfies the Eulerian equations for incompressible perfect fluids, and the gaseous phase surrounding it plays no particular role, apart from by means of the pressure which it exerts. The linearized theory was developed by Rayleigh [RAY 78, LEM 10]. The continuity equation and the irrotationality, as in section 1.3.1.1, yield:

$$\nabla \times \mathbf{v} = \mathbf{0},\ \mathbf{v} = \nabla \phi, \quad \nabla \cdot \mathbf{v} = 0 \rightarrow \nabla^2 \phi = 0 \qquad [1.29]$$

With the cylindrical coordinates being r,θ,x, the velocity vector components are, respectively, v_r,v_θ,v_x. The momentum balance equation is written thus, along the axial coordinate:

$$\frac{\partial v_x}{\partial t}+v_x\frac{\partial v_x}{\partial x}+\frac{1}{\rho}\frac{\partial p}{\partial x}=0 \qquad [1.30]$$

and, after linearization, becomes:

$$\frac{\partial v_x}{\partial t}+\frac{1}{\rho}\frac{\partial p}{\partial x}=0$$

so that:

$$\frac{\partial \phi}{\partial x\partial t}+\frac{1}{\rho}\frac{\partial p}{\partial x}=0$$

The liquid cylinder has a reference radius R and a disturbed radius $R+\eta(x,t)$. In $r=R$, this gives us: $\partial\phi/\partial t+p/\rho=0$, assuming the disturbance on the axis to be null, because of the hypothesis of axial symmetry.

At the liquid–gas interface, we can write the equality of the normal stresses and the impermeability condition:

$$[p]_-^+=\sigma\left(\frac{1}{R_1}+\frac{1}{R_2}\right),\quad \frac{\partial\eta}{\partial t}=\frac{\partial\phi}{\partial r} \qquad [1.31]$$

In the undisturbed reference configuration, the curvature radii are, respectively: $R_1=R, R_2=\infty$ and, in the presence of a small disturbance:

$$R_2=\left[1+(\partial\eta/\partial x)^2\right]^{\frac{3}{2}}\Big/\left(\partial^2\eta/\partial x^2\right)\cong\left(\partial^2\eta/\partial x^2\right)^{-1}$$

The first equation of [1.31] thus becomes, in $r = R$:

$$p \cong -\sigma\left(\eta/R^2 + \partial^2\eta/\partial x^2\right) \qquad [1.32]$$

Thus, we look for solutions of the form: $\phi = X(x)R(r)e^{-i\omega t}$. When substituted into equation [1.29], this formula gives us: $\phi = Ae^{ikx} I_0(kr)e^{-i\omega t}$. The second equation in [1.31] thus gives us: $\eta = \eta_0 e^{ikx} e^{-i\omega t}$.

By injecting these results into the motion equations, we find:

$$\begin{cases} -i\rho A I_0(kr)\omega = \sigma\left(\eta_0/a^2 - k^2\eta_0\right) \\ kA I'_0(kr) = -i\omega\eta_0 \end{cases} \qquad [1.33]$$

System [1.33] then gives us the dispersion equation in $r = a$:

$$\rho\omega^2 = \sigma\left(1/a^2 - k^2\right) k I_1(ka)/I_0(ka) \qquad [1.34]$$

If $ka < 1$, we have a sine curve of increasing amplitude, because ω becomes a pure imaginary number, which corresponds to the instability. We find the maximum amplification factor for: $\omega_{max} = 0.343\sqrt{\sigma/\rho a^3}$, which leads to the wavelength: $\lambda_{max} = 9.02a$.

With this model, we can conceive the formation of identical droplets whose diameter is determined by the following equation for the sphere/cylinder volumes: $\pi d^3/6 = \pi a^2 \lambda$. This gives drops of diameter $d = 1.89(2a)$, which is a little less than two times the diameter of the injector nozzle output.

This corresponds fairly closely to what happens with a vibration injector, by exciting the jet of velocity u at frequency $\nu = u/\lambda_{max}$. We obtain a mono-disperse spray. In practice, this is justified if the wavelength is between 3.5 and 7 times the diameter of the atomizer output, with a velocity greater than the value $4\sigma/\rho a$ [LEM 10].

1.3.2.2. *Linear study of the instability of an isolated viscous liquid cylinder*

The theory is attributed to Weber. For $ka \approx 1$ and after linearization, we obtain the dispersion equation:

$$\rho\omega^2 + 3\mu i\omega k^2 + \sigma\left(1/a^2 - k^2\right) k^2/2 = 0 \qquad [1.35]$$

For the maximum amplification: $1/(ka)_{max} = \sqrt{2(1+3\mu/2a\sigma\rho)}$ and, with the same breakup criterion as in section 1.3.2.1, we obtain the drop diameter:

$$d = 2a\left(\frac{3\pi}{\sqrt{2}}\right)^{1/3}\left(1+\frac{3\mu}{\sqrt{2a\sigma\rho}}\right)^{1/6}$$

The hypothesis $ka \approx 1$ suggests that the drop size is around the same as the radius of the jet which undergoes an in mass disintegration.

1.3.2.3. *Experimental studies of stretched filaments*

The difficulty of observing atomization from liquid filaments in the real conditions of injection into engines has led certain researchers to conduct experiments on liquid bridges [PAD 97, MAR 04a, MAR 04b, VIL 07].

Marmottant and Villermaux [MAR 04b, VIL 04] use a vertical tube submerged into liquid and then rapidly removed. This movement causes that formation of a liquid filament (or ligament) stretched between the surface of the liquid and the bottom end of the tube. The dynamics of the filaments are studied as a function of the elongation rate.

The elongation time t_ε is defined on the basis of the height H of the liquid column, in accordance with the relation:

$$t_\varepsilon = H/(dH/dt) = dt/d \ln H \quad [1.36]$$

It is compared to the capillary time:

$$t_\sigma = \sqrt{\rho_L D^3/\sigma} \quad [1.37]$$

where ρ_L is the density of the liquid, D the diameter of the tube and σ the surface tension between the liquid and gas.

Two cases can be distinguished:

1) slow extension for $t_\sigma/t_\varepsilon << 1$;

2) fast extension for $t_\sigma/t_\varepsilon >> 1$.

In the first case, there is still equilibrium as long as H does not surpass a critical height. Beyond this height, the liquid bridge contracts rapidly, thereby forming a thin cone and a small drop is hence created. The mechanical equilibrium obeys the Laplace law: $\Delta p = \sigma\kappa$ and, for any altitude z, the pressure jump is equal to the hydrostatic pressure, so that: $\sigma\kappa = -\rho g z$. This equation is solved up to a critical height H_{cr} [PAD 73]. We can show that for fairly large tubes: $H_{cr} = 2\ell_\sigma$, where $\ell_\sigma = Dt_g/t_\sigma = \sqrt{\sigma/\rho g}$ is the capillary length, $t_g = \sqrt{D/g}$ being the time taken to fall due to gravity. In the other cases: $H_{cr} = 0.49 D \cong 1.3\ell_\sigma$.

There are, in fact, two contraction regimes depending on the geometry of the flow, which evolves over time, and therefore on the difference from the breakup time t_b (Figure 1.14). In the initial contraction of the whole volume, ξ being the minimum diameter of the column, we have:

$$\frac{\xi}{D} \propto \left(1-\alpha\frac{t}{t_\sigma}\right)^{2/5}$$

When the central part contracts, the motion is self-similar and $\frac{\xi}{D} \propto \left(1-\beta\frac{t}{t_\sigma}\right)^{2/3}$. A similarity is observed for this regime with the case of circular rings connected by the bridges of capillary membranes [CRY 92, CHE 97].

In the second case, a cylindrical column is formed, with the volume $\mathcal{V}_{cyl}$ which is quasi-constant throughout the course of the extension (Figure 1.15). The central part is destabilized, and breaks only after pinching near to the end of the tube or the surface of the tank.

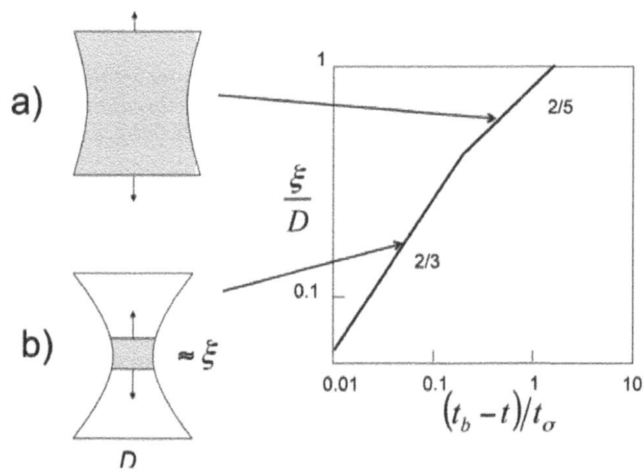

Figure 1.14. *Thickness of the filament at the neck of the tube, as a function of the difference from the breakup time t_b, in the case of slow stretching. Contraction regimes: a) contraction of the whole volume; b) of the central part (figure redrawn from [MAR 04b])*

We shall not go into detail about the Marmottant and Villermaux article [MAR 04a]. Let us simply note that the filament breakup is examined in depth: breakup time, average drop size, models of fragmentation (Kolmogorov model, most likely distribution, aggregations, structure of the filaments during breakup).

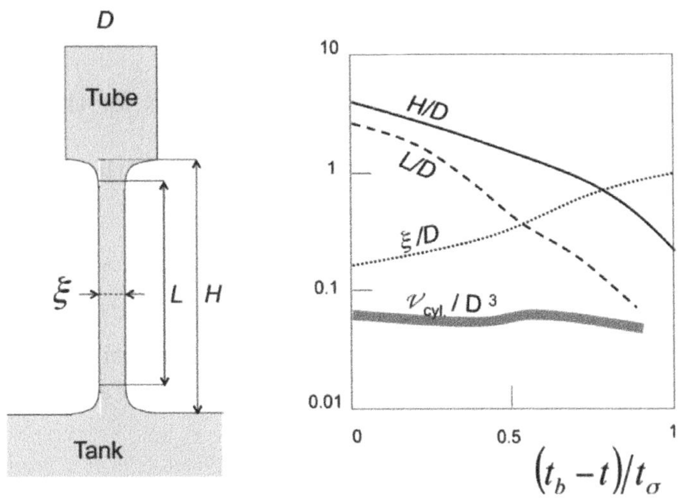

Figure 1.15. *Dimensions of the liquid column over the course of its development as a function of the difference from the breakup time t_b, in the case of fast elongation (figure redrawn from [MAR 04b], with the authors' permission)*

Let us now return to the structure of the filaments. The authors base their thinking on their observations, considering that a ligament consists of a succession of liquid droplets, termed "blobs", which undergo continuous interactions during destabilization. These blobs are divided into sub-blobs, which are themselves arranged in independent adjacent layers along the filament (see Figure 1.16).

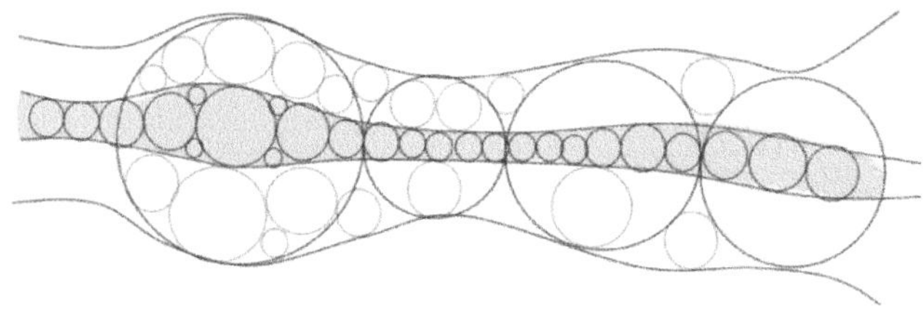

Figure 1.16. *Diagram of the structure of a filament during its destabilization (redrawn from [MAR 04b], with the authors' permission)*

Let $n(d, t)$ be the distribution of the diameter d of the drops, υ the number of independent layers, $q(d',t)$ the distribution of thicknesses d' in each layer formed of sub-blobs, and $\bar{d}$ the average diameter of the blob. As the layers are adjacent, we have: $\bar{d} = \nu \bar{d}'$, and d' is, in a manner of speaking, the mean free path of the liquid motion in the filament.

We use the term:

$$q^{\otimes 2} = \int_{d'=d'_1+d'_2} q(d'_1)\, q(d'_2)\, d(d'_2)$$

to represent the autoconvolution of the distribution $q(d',t)$ [MAR 04b] and set:

$$N = \int_0^\infty n(d,t)\, d(d)$$

The equation of the evolution of $n(d, t)$ found is as follows:

$$\partial n(d,t)/\partial t = -K\, n(d,t)\big(N(t)\big)^{\gamma-1} + \beta K\big(n(d,t)\big)^{\otimes\gamma} \qquad [1.38]$$

Finally, the asymptotic distribution of the blobs obtained is:

$$p\left(x = d/\bar{d}\right) = \frac{\nu^{\nu}}{\Gamma(\nu)} x^{\nu-1} e^{-\nu x} \qquad [1.39]$$

where:

$$\bar{d} = \int d \frac{n(d,t)}{N(t)} d(d)$$

A general remark can be made regarding this article, in which piecemeal solutions are found which accurately explain a variety of behaviors using basic theorems (Bernoulli theorems [PRU 12], log-normal laws[5], etc.). However, no overall solution – numerical or otherwise – is given for the complex problem of atomization by the breakup of liquid filaments. At the present time, no such solution exists.

1.3.3. *Numerical simulation of primary atomization*

Many authors would agree that it is delusional to think it possible to find a complete numerical simulation of primary atomization.

Gorokhovski and Hermann [GOR 08] estimate that the possibilities of solving the question of primary atomization depend not only on the capacity of our computers, but also on the difficulty of modeling it in turbulent flow, taking account of the contribution of the gas–liquid interactions at sub-mesh scale [PRU 13].

---

5 Let X be a log-normal distribution, μ the average (expectation), σ the variance (standard deviation) and Z the standard normal variable. Thus, we have: $X = \exp(\mu + \sigma Z)$. The probability density (mass function) of a log-normal distribution is:

$$P(x;\mu,\sigma) = \frac{1}{\sqrt{2\pi}x\sigma} \exp\left[-\frac{(\ln x - \mu)^2}{2\sigma^2}\right]$$

and we have:

$$\int_0^\infty P(x;\mu,\sigma)\,dx = 1$$

Thus, there is no consensus about the dominant mechanism of breakup when a high-velocity liquid jet makes contact with air at rest. There are multiple phenomena which need to be envisaged: the nonlinear instability of the liquid jet due to the fluctuations in velocity of the flow exiting the injection nozzle, cavitation and collapse of cavitation bubbles, unsteadiness of the injection velocities, etc. A large number of dimensionless parameters are likely to play a part, such as the Weber number (We), the Ohnesorge number (Oh), etc. Therefore, there is a need for phenomenological approaches. Until then, we have numerical approaches such as the RANS (Reynolds-averaged Navier–Stokes) technique.

In spite of the reservations expressed above, DNS manifests definite progress.

1.3.3.1. *Phenomenological approaches such as the Kolmogorov technique*

The approach to primary atomization considered here (initially applied to secondary atomization) combines LES (Large Eddy Simulation) in the turbulent gaseous flow with stochastic simulation of the scattering of the liquid jet. The frequency of breakup is assumed to be sufficiently high to consider the destruction of the liquid jet as a *cascaded fragmentation* with scale symmetry.

The parameters of the stochastic process are the ratio characterizing the spectrum of drops produced in the vicinity of the injector and the kinetic energy of the transfer from the gas to the liquid.

According to Kolmogorov, each break of a fragment reduces its size r by a factor α (which becomes αr), governed by a fragmentation spectrum $q(\alpha)$ such as $\int_0^1 q(\alpha)d\alpha = 1$.

Let ν be the frequency of fragmentation, and $f(r,t)$ the normalized distribution function. We find the following equation for the kinetics of fragmentation:

$$\frac{1}{\nu}\frac{\partial f(r,t)}{\partial t}=\int_0^1 \frac{1}{\alpha} f\left(\frac{r}{\alpha},t\right) q(\alpha)\,d\alpha - f(r,t) \qquad [1.40]$$

The steady-state solution to this equation (we make $\partial/\partial t = 0$ in equation [1.40]) is the Dirac distribution:

$$f(r,t)=\delta(r) \qquad [1.41]$$

A priori, we do not know how to reach this steady-state solution if we do not know the value of $q(\alpha)$.

However, because of the scale symmetry, $f(r,t)$ passes through two asymptotic distributions. In order to determine these distributions, we do not need to determine the spectrum $q(\alpha)$ – merely its first two logarithmic moments. Additionally, when the time is long, the ratio between these moments $\langle \ln\alpha\rangle / \langle \ln^2\alpha\rangle$ entirely determines the solution to equation [1.40]. Gorokhovski and Saveliev [GOR 03] gave the two universally-applicable formulas:

$$r\,f(r,t)=\frac{1}{\sqrt{2\pi\langle \ln^2\alpha\rangle \nu t}}\, e^{-\frac{\langle \ln\alpha\rangle^2}{2\langle \ln^2\alpha\rangle}\nu t}\, e^{-\frac{(\ln r/R)^2}{2\langle \ln^2\alpha\rangle \nu t}} \left(\frac{r}{R}\right)^{\frac{\langle \ln\alpha\rangle}{\langle \ln^2\alpha\rangle}} \qquad [1.42]$$

where $R=e^{\langle \ln\alpha\rangle_0}$ and $\left\langle \ln^l \alpha\right\rangle = \int_0^1 \ln^l \alpha\, q(\alpha)\,d\alpha$.

We can show that:

$$r\,f(r,t\to\infty)\propto \left(\frac{r}{R}\right)^{\frac{\langle \ln\alpha\rangle}{\langle \ln^2\alpha\rangle}} \qquad [1.43]$$

As we can demonstrate that:

$$\left\langle \left(\ln r - \langle \ln r \rangle\right)^2 \right\rangle / \langle \ln r \rangle = \langle \ln^2 \alpha \rangle / \langle \ln \alpha \rangle$$

we find that, at high frequencies, equation [1.40] becomes:

$$\frac{1}{\nu}\frac{\partial f(r,t)}{\partial t} = -\langle \ln \alpha \rangle \frac{\partial r f(r)}{\partial r} + \frac{1}{2}\langle \ln^2 \alpha \rangle \frac{\partial}{\partial r}\left(r \frac{\partial r f(r)}{\partial r} \right) \qquad [1.44]$$

This is a Fokker–Planck equation describing the stochastic log-Brownian process:

$$\dot{r} = \nu \langle \ln \alpha \rangle r + \sqrt{\nu \langle \ln^2 \alpha \rangle / 2 r \Gamma(t)} \qquad [1.45]$$

with $\Gamma(t)$ being the Langevin process such that:

$$\langle \Gamma(t) \rangle = 0, \text{et } \langle \Gamma(t) \Gamma(t') \rangle = 2\delta(t'-t)$$

1.3.3.2. *RANS-type approaches to primary atomization*

Turbulence in a two-phase mixture has long been a topic of study [TCH 47]. Borghi *et al.* propose the RANS approach for a two-phase mixture in which the liquid and gas are moving independently [VAL 01].

"The modeling of the atomization of a liquid jet in a gaseous environment is based on the principle that in the very dense zone near to the injector nozzle, a Lagrangian approach is not appropriate, because of the presence of the liquid core and the significant interactions between the liquid and gas flows. Thus, an Eulerian mixing approach, which considers the liquid and gaseous phases as a complex mixture and represents them as a single fluid with highly variable density, seems more appropriate to describe the dense zone of the spray [BEA 06]. Later on, once the flow becomes sufficiently dilute, we can model the two-phase flow using Lagrangian models".

Blokkeel [BLO 03] proposed to couple the Eulerian representation of the dense zone of a spray with a Lagrangian calculation in the dilute zone. This idea gave rise to the ELSA mode (for *Eulerian Lagrangian Spray Atomization*), developed to represent the atomization and evolution of a spray from the dense zone to the dilute zone.

Inter-phase mixing is thus assimilated to turbulent mixing, even though the (two-phase) turbulence actually encountered is a very particular case. The choice of the mixing model also bypasses the complex handling of the interface transfer terms. It can be legitimately suggested that in the dense zone of the spray, the flow appears to be more a mixture of liquid and gas (from a macroscopic point of view) than two truly separate phases.

Thus, for a diesel engine, this model constitutes a complete method for description of the jet from the inside of the injector to the dilute zone of the spray.

This ELSA model is based on two hypotheses:

– the Weber and Reynolds numbers are assumed to be large (which is generally the case for diesel direct injection engines);

– the two-phase mixture (liquid and gas) is studied as a single fluid with a highly variable density.

Here, we shall limit our examination to the Eulerian part of the code, presenting the basic equations inspired by the $k-\varepsilon$ method for turbulent flows of compressible mixtures [PRU 13].

We define the average density of the mixture by its inverse:

$$\frac{1}{\bar{\rho}} = \frac{\tilde{Y}}{\rho_l} + \frac{1-\tilde{Y}}{\rho_g} \qquad [1.46]$$

where ρ_l, ρ_g are the densities, and $\tilde{Y}$ is the Favre average of the liquid fraction.

The balance equation of the mass fraction of liquid (if we discount evaporation in the dense zone) becomes:

$$\frac{\partial \bar{\rho}\tilde{Y}}{\partial t} + \nabla \cdot \left(\bar{\rho}\tilde{\mathbf{v}}\tilde{Y}\right) = -\nabla \cdot \overline{\rho \mathbf{v}''Y''} \qquad [1.47]$$

with $\tilde{\mathbf{v}}$ being the Favre average of the velocity of the mixture. The right-hand side of the equation describes the dispersion of the liquid phase by turbulence.

The momentum equation for the mixture is written thus:

$$\frac{\partial \bar{\rho}\tilde{\mathbf{v}}}{\partial t} + \nabla \cdot \left(\bar{\rho}\tilde{\mathbf{v}} \otimes \tilde{\mathbf{v}}\right) = -\nabla\bar{p} - \nabla \cdot \overline{\rho \mathbf{v}'' \otimes \mathbf{v}''} \qquad [1.48]$$

This equation is classic for turbulent gaseous mixtures, but here the variations in density are much greater than usual. The Reynolds stresses on the right-hand side are written as follows, if we accept the Boussinesq approximation combined with the $k-\varepsilon$ model:

$$\overline{\rho \mathbf{v}'' \otimes \mathbf{v}''} = -\bar{\rho}\nu_t \tilde{\mathbf{D}} + \frac{2}{3}\bar{\rho}\tilde{k}\mathbf{1} \qquad [1.49]$$

where $\tilde{k}$ is the Favre average of the mean turbulent kinetic energy of the mixture, $\tilde{\varepsilon}$ is the average turbulent dissipation and ν_t is the turbulent kinematic viscosity, equal to:

$$\nu_t = C_\mu \frac{\tilde{k}^2}{\tilde{\varepsilon}} \text{ and } C_\mu = 0.09$$

The turbulent kinetic energies of the liquid, gas and the mixture, respectively, are:

$$\overline{k}_l = \frac{1}{2}\overline{\mathbf{v}_l' \cdot \mathbf{v}_l'}, \quad \overline{k}_g = \frac{1}{2}\overline{\mathbf{v}_g' \cdot \mathbf{v}_g'}, \quad \tilde{k} = \frac{1}{2\overline{\rho}}\overline{\rho\mathbf{v}'' \cdot \mathbf{v}''} \qquad [1.50]$$

and we have:

$$\overline{\rho}\tilde{k} = \overline{\rho}\tilde{Y}\overline{k}_l + \overline{\rho}\left(1-\tilde{Y}\right)\overline{k}_g + \frac{1}{2\overline{\rho}}\overline{\rho\mathbf{v}''Y''}^2\left(\frac{1}{\tilde{Y}} + \frac{1}{1-\tilde{Y}}\right) \qquad [1.51]$$

Similarly, the respective dissipation rates are written as:

$$\overline{\varepsilon}_l = 2\nu_t\overline{\nabla\otimes\mathbf{v}_l' : \nabla\otimes\mathbf{v}_l'},\ \overline{\varepsilon}_g = 2\nu_t\overline{\nabla\otimes\mathbf{v}_g' : \nabla\otimes\mathbf{v}_g'},\ \tilde{\varepsilon} = 2\nu_t\overline{\nabla\otimes\mathbf{v}'' : \nabla\otimes\mathbf{v}''} \qquad [1.52]$$

and we have:

$$\tilde{\varepsilon} = \tilde{Y}\overline{\varepsilon}_l + \left(1-\tilde{Y}\right)\overline{\varepsilon}_g + \frac{2\nu_t}{\overline{\rho}^2}\vec{\nabla}\otimes\overline{\rho\mathbf{v}''Y''} : \vec{\nabla}\otimes\overline{\rho\mathbf{v}''Y''}\left(\frac{1}{\tilde{Y}} + \frac{1}{1-\tilde{Y}}\right) \qquad [1.53]$$

The balance equations satisfied by the kinetic energy and the turbulent dissipation rates are similar to those of a single-phase environment [DEM 06].

We can also establish [BEA 06] the transport equation for the average density of the liquid/gas interface area [VAL 01], as has been done for flames with the coherent flame model (see [MAR 77, TRO 94]; section 3.3.4.5 in [PRU 13] and section 8.4.4 – equation [8.127] – in [PRU 10a]).

Mixture density	$\frac{1}{\overline{\rho}}=\frac{\widetilde{Y}}{\rho_l}+\frac{1-\widetilde{Y}}{\rho_g}$
Momentum	$\frac{\partial\overline{\rho}\widetilde{U}_i}{\partial t}+\frac{\partial\overline{\rho}\widetilde{U}_j\widetilde{U}_i}{\partial x_j}=-\frac{\partial\overline{P}}{\partial x_i}-\frac{\partial\overline{\rho u_i''u_j''}}{\partial x_j}$
Mass fraction of liquid	$\frac{\partial\overline{\rho}\widetilde{Y}}{\partial t}+\frac{\partial\overline{\rho}\widetilde{U}_j\widetilde{Y}}{\partial x_j}=\frac{\partial}{\partial x_j}\overline{\rho}\frac{\nu_t}{Sc_t}\frac{\partial\widetilde{Y}}{\partial x_j}$
QMM two-phase model	$\frac{\partial\overline{\rho u_i''y''}}{\partial t}+\frac{\partial\widetilde{U}_j\overline{\rho u_i''y''}}{\partial x_j}=-\frac{\partial}{\partial x_j}\left(\overline{\rho}\frac{\nu_t}{Sc_t}\frac{\partial}{\partial x_j}\frac{\overline{\rho u_i''y''}}{\overline{\rho}}\right)-C_{f1}\overline{\rho u_j''y''}\frac{\partial\widetilde{U}_i}{\partial x_j}$ $-C_{f2}\overline{\rho u_i''u_j''}\frac{\partial\widetilde{Y}}{\partial x_j}-\frac{C_{f3}}{\left(1-\widetilde{Y}\right)\tau_p}\left(\overline{\rho u_j''y''}+\overline{\rho}D_{gl,t}\frac{\partial\widetilde{Y}}{\partial x_i}\right)$
QMM single-phase model	$\frac{\partial\overline{\rho u_i''y''}}{\partial t}+\frac{\partial\widetilde{U}_j\overline{\rho u_i''y''}}{\partial x_j}=-\frac{\partial}{\partial x_j}\left(\overline{\rho}\frac{\nu_t}{Sc_t}\frac{\partial}{\partial x_j}\frac{\overline{\rho u_i''y''}}{\overline{\rho}}\right)-C_{f1}\overline{\rho u_j''y''}\frac{\partial\widetilde{U}_i}{\partial x_j}$ $-C_{f2}\overline{\rho u_i''u_j''}\frac{\partial\widetilde{Y}}{\partial x_j}-C_{f3}\frac{\overline{\rho u_i''y''}}{\tau_t}$
Liquid/gas interfacial density	$\frac{\partial\overline{\rho}\widetilde{\Omega}}{\partial t}+\frac{\partial\overline{\rho}\widetilde{\Omega}\widetilde{U}j}{\partial x_j}=\frac{\partial}{\partial x_j}\left(\overline{\rho}\frac{\nu_t}{Sc_t}\frac{\partial\widetilde{\Omega}}{\partial x_j}\right)+\overline{\rho}\left(C_{\Omega1}\dot{\tilde{\Omega}}_{mixing}+C_{\Omega2}\dot{\tilde{\Omega}}_{turbulent\ stretching}\right)$ $+\overline{\rho}\left(C_{\Omega3}\dot{\tilde{\Omega}}_{average\ stretching}+C_{\Omega4}\dot{\tilde{\Omega}}_{breakup}+C_{\Omega5}\dot{\tilde{\Omega}}_{coalescence}\right)$

Table 1.1. *Recap of the equations solved in the ELSA model; the acronym "QMM" means Quasi-Multiphasic Model (drawn from [BEA 06])*

We use the area per unit mass $\tilde{\Omega}=\overline{\Sigma}/\overline{\rho}$ rather than the average area per unit volume $\overline{\Sigma}$. The balance relation for the density of interface area is written as:

$$\frac{\partial\left(\overline{\rho}\,\tilde{\Omega}\right)}{\partial t}+\nabla\cdot\left(\overline{\rho}\,\tilde{\Omega}\otimes\tilde{\mathbf{v}}\right)=\nabla\cdot\left(\overline{\rho}\frac{\nu_t}{Sc_t}\nabla\tilde{\Omega}\right)+$$
$$\overline{\rho}\left(C_{\Omega1}\,\dot{\tilde{\Omega}}_{mixing}+C_{\Omega2}\,\dot{\tilde{\Omega}}_{turbulent\ stretching}\right)+\qquad[1.54]$$
$$\overline{\rho}\left(C_{\Omega3}\,\dot{\tilde{\Omega}}_{average\ stretching}+C_{\Omega4}\,\dot{\tilde{\Omega}}_{breakup}+C_{\Omega5}\,\dot{\tilde{\Omega}}_{coalescence}\right)$$

The equations solved in the ELSA model are given in Table 1.1.

1.3.3.3. *Direct Numerical Simulation (DNS)*

Many researchers have tried to tackle the direct numerical simulation of the problem of breakup of liquids in gases. The team led by S. Zaleski at the Institut Dalembert at France's UPMC and the CNRS has focused on the simulation of the motion and generation of drops and bubbles, and on the process of atomization [YEC 05, TRY 11] (see figure on the Institut Dalembert Website[6]).

The main difficulties encountered relate to the simulation of the liquid–gas interfaces with tracking of the surfaces in the computational mesh, taking account of the surface tensions and gravity.

The numerical method developed here is that of volumes of fluid (VOF), and the efforts in this field have focused particularly on the reconstruction of interfaces from discrete segments and their propagation using a Lagrangian method.

Let us look at the article by Agbaglah *et al.* [AGB 11] on adaptive parallel octree simulation of multi-phase flows by monitoring of the interface of the representative volume of fluid. This article describes simulations performed with the Gerris code – a free software program which implements finite-volume resolution methods on an adaptive hierarchical octree mesh, and a method for volume tracking with affine piecewise construction of the interface. The Gerris parallelization is obtained by decomposition of the space into different zones. Examples of the capabilities of this approach are shown on many types of problems. The impact of a drop onto a layer of the same liquid causes the formation of a thin

6 http://www.lmm.jussieu.fr/~zaleski/drops.html.

layer of air under the drop at the time of impact, which can be captured using the adaptive method. It is followed by the shooting up of a thin corolla above the impacting drop. The problem of atomization is also a significant challenge for intensive calculus, in which a large number of small-scale structures are produced.

Finally, the authors show an example of turbulent jet computation in an equivalent resolution of 6×1024^3 cells.[7]

The figure presented by Agbaglah *et al.* [ABG 11] contains a three-dimensional representation of the oil/water interface for Fr = 10. Taking the diameter of the pipe, D, as a reference distance, a $40 \times 60 \times 20$ domain is simulated, with a minimum mesh of 0.0195. The domain is first constructed with 6×20^3 cubic boxes; these boxes are then subdivided into 10 levels of refinement, giving the equivalent of a 6×1024^3 resolution.

Two criteria are used to dynamically adapt the mesh: the vorticity norm and an *a posteriori* error estimation on the prediction of the concentration field. These criteria are used depending on the cost of the adaptation. The *a posteriori* error is estimated by calculating the norm of the Hessian matrix of the given domain, using the discretization operators with 3rd-order precision. The use of adaptive techniques reduces the number of cells by three orders of magnitude in relation to the equivalent Cartesian-space simulation, which makes it possible to perform this type of simulations in a 32-processor cluster in 50 hours of CPU time (the duration of the simulation, in seconds, is: $35.7\sqrt{D/g}$, where D is the diameter of the pipe and g is the acceleration due to gravity).

7 The jet simulation is based on the configuration of the Deepwater Horizon oil leak.

The VOF method and the numerical codes associated therewith, such as Gerris, have demonstrated their effectiveness in numerous cases. The problems of capillary action and the presence of gravity have been solved. They lend themselves well in adaptation to real-world flows comprising a reactive, compressible multi-species gaseous phase.

1.3.4. *Secondary atomization*

The destabilization of the film and then of the layer, leading to the formation of droplets (section 1.3.1), possibly by way of the formation of filaments (see section 1.3.2), is known as the primary breakup mechanism or primary atomization.

Once the mist of drops has been formed, it is then subjected to the acceleration of the gaseous flow, which can cause phenomena such as the coalescence of drops or indeed the breakup of large drops into smaller ones. These mechanisms are known as secondary atomization.

Thus, the final distribution of the sizes of the drops depends simultaneously on the sizes of the drops produced during primary atomization and the strength of the secondary mechanisms [LAL 06].

In other words, the final distribution of the drop sizes depends on the situation created by the primary atomization.

The Weber number (see Table A1.2 in [PRU 13]), constructed from the relative velocity between the gas and the liquid drop: $We = \rho(V_G - V_L)^2 D/\sigma$, ultimately determines what happens.

1.3.4.1. *Drops suddenly subjected to a gaseous flow*

If we place drops in a fast gaseous flow, these drops are greatly deformed and break up into multiple droplets, which may, in turn, come together and agglomerate.

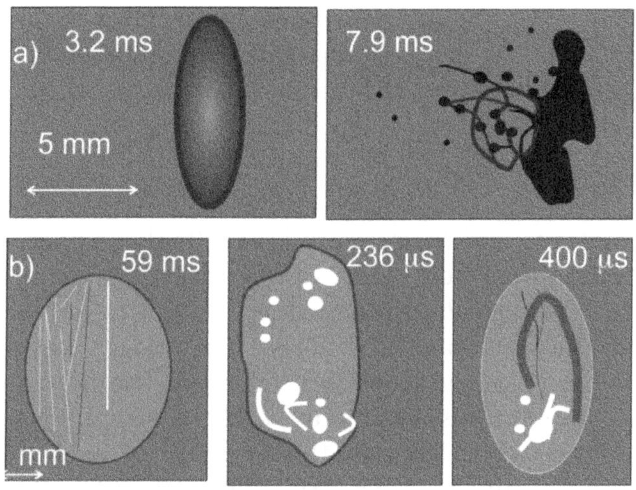

Figure 1.17. *Diagram of the evolution of a drop suddenly subjected to a gaseous flow, captured at different times. a) We = 5.4 × 10³; b) We = 75 (figure redrawn from the photos in Theofanous and Li [THE 08])*

One of the earliest and most cited articles on the subject is that of Hanson and Domich [HAN 63], where the authors place drops in the path of a blastwave.

More recently, Theofanous and Li [THE 08] experimentally examined "aerodynamic breakup". They used shadowgraph with laser fluorescence to observe liquid drops suddenly exposed to supersonic gaseous flows. They note that with high Weber numbers $We > 10^3$, the dominant mechanism is the shear motion induced, with a large radial component and instabilities in the stretched liquid layer thus generated (see Figure 1.17(a)). For low Weber numbers $We < 10^2$, the results show numerous multi-wave drillings by Rayleigh–Taylor instabilities (Figure 1.17(b)).

Villermaux and Bossa [VIL 09], in their study of the formation of raindrops, show that, whilst the initial growth of the drops in the clouds results from interactions and coalescence between neighboring droplets, the final distribution of the drops in free-fall tends to stem more from internal fragmentation of large isolated drops which become deformed as they fall through the air.

1.3.4.2. *Drops formed from concentric jets*

Marmottant and Villermaux [MAR 04a] consider a liquid jet entering into a flow of air. In this axisymmetric case, they estimate that secondary atomization cannot take place. This is due to the fact that the difference in velocity between the gas and the primary drops forming from the filaments (see section 1.3.2) diminishes rapidly, and the size distribution of the droplets (which follows the breakup of each filament) is essentially fixed [MAR 01]. The size distribution of the drops is therefore the sum of those resulting from the breakup of each filament. The largest drops tend to relax into a spherical form, and we see the dilution of the spray in the surrounding gas, rather than experiencing a series of new breakups.

The measurements show a law in terms of $p(d) \approx \exp\left(-nd/\langle d_0 \rangle\right)$, where $n \approx 3.5$, increasing gradually with the velocity of the gas.

With regard to the breakup of the filaments into drops, remember the reference [MAR 04b]. The maximum entropy method (see a summary in [PRU 10a]) is harshly criticized in [VIL 07], because it is founded on the idea of a stochastic distribution of the droplets in sprays, whereas Villermaux takes as his starting point a distribution deduced from statistical thermodynamics applied within each filament, as we saw in section 1.3.2.3.

Finally, let us point to the interesting studies on atomization conducted by Sellens and Zeng [SEL 98], Lasheras, Villermaux and Hopfinger [LAS 98] and more recently, Planchette, Laurenceau and Brenn [PLA 12].

2

Problems at the Scale of a Particle

In Chapter 3, regarding flows of suspensions, we shall see that the process involves a force $\vec{\mathcal{F}}$ and a heat flux $\mathcal{Q}$ characterizing the mechanical and thermodynamic actions of the "fluid phase" on the "particle phase" per unit volume of the mixture. Linearized TIP (see Volume 1 [PRU 12]) provides us with expressions for these values, deduced from phenomenological laws. Two characteristic times, τ_T and τ_V, come into play.

In this chapter, we present the theoretical analyses used to determine these characteristic times, by examining the flows at the mesoscopic scale – i.e. at the scale of each particle. The theory yields expressions which are generally corrected by empirical formulas.

If we step away a little from the simplified hypothesis of a single-component fluid and neutral particles, we are faced with the presence of numerous phenomena. One such phenomenon is the variation of mass of the particles due to the influence of evaporation and condensation. The case of a drop in the presence of a diffusion flame will be examined. In each of these problems, the particle will be assumed to be spherical.

With regard to the appearance of a bubble of vapor (or dissolved gas precipitating out) in a liquid, its motion and the phenomenon of collapse, readers could usefully consult [CHA 13]. In addition, various authors have focused on the behavior of particles at the fluid interfaces [KRA 01].

2.1. Force exerted by a fluid on a spherical particle

To begin with, the fluid will be taken to be incompressible, and various situations will be studied before arriving at a realistic formula [GER 62, FOR 67].

2.1.1. *Perfect incompressible fluid*

A sphere placed in a steady-state flow of perfect fluid with irrotational motion with velocity at infinity U_∞ is subject only to the Archimedes forces. The motion of the fluid induces no other force. If R is the radius of the sphere, the force is then directed along the ascending vertical, and has the value:

$$F = \frac{4}{3}\pi R^3 \rho_{gs} \qquad [2.1]$$

The flow can be described with the stream function Ψ or the velocity potential ϕ, both of which are harmonics. The stream function satisfies the equation:

$$\frac{\partial^2 \psi}{\partial x^2} + \frac{\partial^2 \psi}{\partial y^2} - \frac{1}{y}\frac{\partial \psi}{\partial y} = 0 \qquad [2.2]$$

It results from the superposition of a doublet (see section 1.5.2.3 of Volume 1, [PRU 12]) of intensity: $K = -2\pi U_\infty R^3$, which gives us:

$$\psi = \frac{K}{4\pi}\frac{\sin^2\theta}{r} = \frac{K}{4\pi}\frac{y^2}{r^3} \qquad [2.3]$$

and a uniform flow giving us:

$$\psi = U_\infty \frac{r^2 \sin^2\theta}{2} = U_\infty \frac{y^2}{2} \qquad [2.4]$$

Thus, we obtain:

$$\phi = U_\infty x - \frac{K}{4\pi}\frac{x}{r^3},\ \psi = \frac{U_\infty}{2} y^2 + \frac{K}{4\pi}\frac{y^2}{r^3} \qquad [2.5]$$

The variables x, y and r are defined by Figure 2.1. This solution is valid only in the case of cylindrical symmetry around the Ox axis.

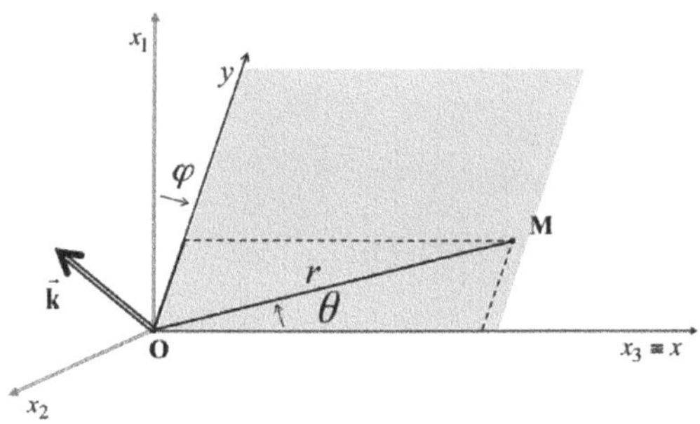

Figure 2.1. *Configuration with cylindrical symmetric and spherical coordinates* r,θ,φ

If, instead of being immobile in a uniform infinite flow, the sphere has a uniform translational motion of velocity $\mathbf{v}_A$ in an atmosphere at rest at infinity, the result is the same in a frame of reference linked to the sphere. If $\mathbf{v}_A$ is not constant, the flow field is determined in a similar manner at any given time, but the intensity of the doublet is variable over time. This gives rise to a force stemming from the sphere's acceleration. This force, which is proportional to the acceleration, brings an induced mass into play. To obtain its value, we need to switch from the system of axes in which the sphere's motion is observed to the system of axes connected to the sphere itself (see Figure 2.2).

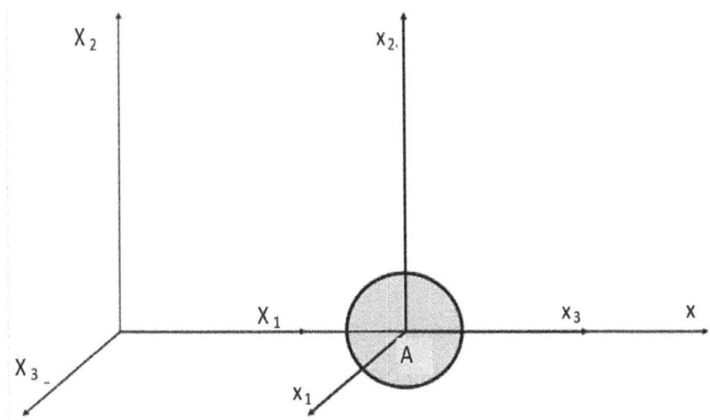

Figure 2.2. *System of axes connected to the sphere*

Thus, we have:

$$\phi(X_1, X_2, X_3, t) = \varphi(x_1, x_2, x_3, t) \quad [2.6]$$

We can write that at any given point on the surface, the relative normal velocity of the flow is null – i.e.:

$$(\mathbf{v} - \mathbf{V}_A)\cdot\mathbf{n} = 0 \quad [2.7]$$

or indeed:

$$\partial\varphi/\partial r = V_A \cos\theta \quad [2.8]$$

The potential $\varphi = -R^3 V_A \cos\theta / 2r^2$ satisfies this condition.

Thus, for the potential ϕ, we obtain the function:

$$\phi = -R^3 (\mathbf{V}_A \cdot \mathbf{AM}) / 2AM^3 \quad [2.9]$$

Note that: $\partial \mathbf{AM}/\partial t = -\mathbf{V}_A$. The application of the unsteady Bernoulli equation enables us to calculate the forces exerted on the obstacle. Only the term $\partial\phi/\partial t$ gives a non-null result, aside from the Archimedes force mentioned above.

We obtain:

$$\frac{\partial \phi}{\partial t} = -\frac{R^2}{2}\left(\frac{\gamma_A \cdot \mathbf{AM}}{r^3} - \frac{|\mathbf{V}_A|^2}{r^3} + 3\frac{\mathbf{V}_A \cdot \mathbf{AM}}{r^5}\right) \qquad [2.10]$$

The last two terms disappear from the expression of the added mass force for reasons of symmetry, and we are left with:

$$\mathbf{F}' = -\frac{\rho}{2r}\int_{\Sigma}(\gamma_A \cdot \mathbf{AM})\mathbf{AM}\, ds = -\frac{2}{3}\pi \rho R^3 \gamma_A = -m'\gamma_A \qquad [2.11]$$

The induced mass m' is equal to half the mass that the sphere would have if it were filled with the surrounding fluid. The induced force of inertia $\vec{\mathbf{F}}'$ also exists in the presence of viscous fluid, but in that case its expression is different.

2.1.2. *Incompressible viscous fluid*

2.1.2.1. *General points*

The particles of a two-phase flow are driven by the carrier fluid, primarily under the influence of the forces of viscous friction. In most cases, the Reynolds number, based on the relative velocities of the gas and particles, and their radius, or their diameter, is very small. In the vicinity of the particle, the forces of inertia of the fluid are therefore negligible in comparison to the forces of viscosity, so that the nonlinear terms in the momentum balance equation are discounted. Therefore, an initial approximation of the system describing the steady-state flow is:

$$\nabla \cdot \mathbf{v} = 0, \quad \nabla p = \mu \nabla^2 \mathbf{v} \tag{2.12}$$

These equations are at the root of the *Stokes' theorem*, which is capable of correctly calculating the force of friction. The flow obtained, however, is not very satisfactory, in that, firstly, the stream lines remain highly disturbed far away from the sphere, and secondly, no wake is obtained, which runs counter to the experimental observations. In order to obtain a better description far from the sphere, we use *Oseen's theory*, which takes account of the convection induced by the velocity at infinity. If that velocity is $\mathbf{U}_\infty$, then we have the system:

$$\nabla \cdot \mathbf{v} = 0, \; \rho \mathbf{U}_\infty \cdot \nabla \otimes \mathbf{v} + \nabla p = \mu \nabla^2 \mathbf{v} \tag{2.13}$$

which is tantamount to linearizing the inertial terms.

Let us first establish the complete system of equations in the general case, without overlooking the nonlinear terms, and in the case of a flow of revolution. The variables used are those from the previous section. In order to eliminate the pressure we consider the curl of both sides of the momentum balance equation, which becomes:

$$\frac{\partial \boldsymbol{\omega}}{\partial t} + \nabla \times (\boldsymbol{\omega} \times \mathbf{v}) = \nu \nabla^2 \boldsymbol{\omega} \tag{2.14}$$

where $\nu = \dfrac{\mu}{\rho}$ and $\vec{\omega} = \dfrac{1}{2} \vec{\nabla} \times \vec{v}$. The continuity equation becomes:

$$\frac{\partial (yu)}{\partial x} + \frac{\partial (yv)}{\partial y} = 0$$

which gives us the stream function ψ such that: $u = \frac{1}{y}\frac{\partial \psi}{\partial y}$, $v = -\frac{1}{y}\frac{\partial \psi}{\partial x}$.

We can show that the vector **ω** is carried by the vector **k** which, with **i** and **j**, forms a direct trirectangular trihedron (see Figure 2.3), and that its coordinate in the direction Oz is:

$$\omega = -\frac{1}{2y}\left(\frac{\partial^2 \psi}{\partial x^2} + \frac{\partial^2 \psi}{\partial y^2} - \frac{1}{y}\frac{\partial \psi}{\partial y}\right) \quad [2.15]$$

The ω equation thus becomes:

$$\frac{\partial \omega}{\partial t} + u\frac{\partial \omega}{\partial x} + v\frac{\partial \omega}{\partial y} - \frac{\omega v}{y} = -\frac{\nu}{y}\left(\frac{\partial^2(\omega y)}{\partial x^2} + \frac{\partial^2(\omega y)}{\partial y^2} - \frac{1}{y}\frac{\partial(\omega y)}{\partial y}\right) \quad [2.16]$$

The switch to dimensionless values brings the Reynolds number $Re = U_\infty R/\nu$ into play, when we take the following reference values: time R/U_∞, velocity U_∞, length R, stream function $U_\infty R^2$, eddy U_∞/R.

Using these last two equations ([2.15] and [2.16]) it is possible to study the flow.

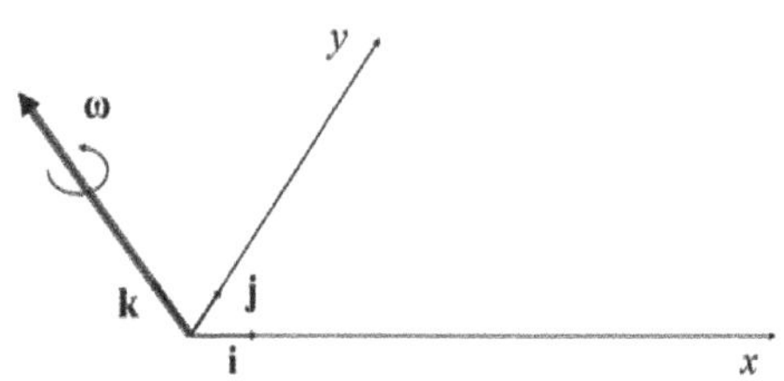

Figure 2.3. *Trirectangular trihedron in space*

Without changing the notations, the above equations give us the following dimensionless system:

$$\begin{cases} \dfrac{\partial^2\psi}{\partial x^2}+\dfrac{\partial^2\psi}{\partial y^2}-\dfrac{1}{y}\dfrac{\partial\psi}{\partial y}=-2\omega y \\ \dfrac{\partial^2(\omega y)}{\partial x^2}+\dfrac{\partial^2(\omega y)}{\partial y^2}-\dfrac{1}{y}\dfrac{\partial(\omega y)}{\partial y}=\mathrm{Re}\, y\left(\dfrac{\partial\omega}{\partial t}+u\dfrac{\partial\omega}{\partial x}+v\dfrac{\partial\omega}{\partial y}-\dfrac{\omega v}{y}\right) \end{cases} \qquad [2.17]$$

2.1.2.2. *Stokes' theorem*

In the case of the *Stokes' theorem*, the convective terms on the right-hand side of the second equation in [2.17] disappear ($\mathrm{Re}=0$), and it becomes:

$$\frac{\partial^2(\omega y)}{\partial x^2}+\frac{\partial^2(\omega y)}{\partial y^2}-\frac{1}{y}\frac{\partial(\omega y)}{\partial y}=0 \qquad [2.18]$$

This is the same equation as [2.2], satisfied by the stream function of an irrotational incompressible perfect fluid: ψ is simply replaced by (ωy). Thus, we are already aware of one particular solution – that of the doublet of the form $\omega y = K\sin^2\theta/2r$, which we shall adopt here. Equation [2.2] then gives us:

$$\frac{\partial^2\psi}{\partial r^2}+\frac{1}{r^2}\frac{\partial^2\psi}{\partial\theta^2}-\frac{\cos\theta}{r^2\sin^2\theta}\frac{\partial\psi}{\partial\theta}=-\frac{K\sin^2\theta}{r} \qquad [2.19]$$

By setting $\psi=f(r)\sin^2\theta$, we obtain:

$$f''(r)-\frac{2}{r^2}f(r)=-\frac{K}{r} \qquad [2.20]$$

The resolution is immediate. In view of the boundary conditions at the sphere where $\boldsymbol{v}$ = 0: $\psi(1,\theta)=0, \dfrac{\partial\psi(1,\theta)}{\partial r}=0$ and the conditions at infinity where $\psi=y^2/2$, we obtain the value $K = -3/2$ for the intensity of the doublet and the solution:

$$\psi = \frac{\sin^2 \theta}{2}\left(r^2 - \frac{3r}{2} + \frac{1}{2r}\right) \qquad [2.21]$$

The r^2 and $1/2r$ terms are identical to those of an irrotational flow of perfect fluid: i.e. the superposition of a uniform flow and a doublet. The term $-(3/4)\sin^2 \theta$ is specific to this case. It is sometimes called a *Stokelet*.

As the velocity field is known, we can calculate the forces acting on the sphere, e.g. by writing that their power is equal to the power dissipated because of the viscosity: $U_\infty F = 4\mu \int_V \omega^2 dV$. We obtain the *Stokes drag* force (in dimensional variables):

$$F = 6\pi \mu U_\infty R \qquad [2.22]$$

The force per unit mass is therefore: $\frac{F}{P_p} = \frac{F}{4/3\pi R^3 Pps} = \frac{9\mu U\infty}{2y^2 Pps}$. Thus, if we replace $U\infty$ with the local difference between the macroscopic velocity of the gas and that of the particles $(v_g - v_p)$, we find:

$$\mathcal{F} = \rho_p \frac{\mathbf{v}_g - \mathbf{v}_p}{\tau_v} \qquad [2.23]$$

where $\tau_v = 2R^2 \rho_{ps}/9\mu$ is the expression of the friction relaxation time given by Stokes' theorem.

2.1.2.3. *Oseen theory*

In the *Oseen theory*, the dimensionless system needing to be solved becomes:

$$\begin{cases} \dfrac{\partial^2 \psi}{\partial x^2} + \dfrac{\partial^2 \psi}{\partial y^2} - \dfrac{1}{y}\dfrac{\partial \psi}{\partial y} = -2\omega y \\ \dfrac{\partial^2 (\omega y)}{\partial x^2} + \dfrac{\partial^2 (\omega y)}{\partial y^2} - \dfrac{1}{y}\dfrac{\partial (\omega y)}{\partial y} = \mathrm{Re}\, y \left(u \dfrac{\partial \omega}{\partial x} \right) \end{cases} \quad [2.24]$$

The Oseen solution for the sphere is:

$$\psi = \left(r^2 + \frac{1}{2r} \right) \frac{\sin^2 \theta}{2} - \frac{3}{2} \frac{1 + \cos\theta}{\mathrm{Re}} \left[1 - \exp(- \frac{\mathrm{Re}(1-\cos\theta) r}{2} \right] \quad [2.25]$$

This contains the terms from the uniform flow [MAX 83], from the doublet, and one more case-specific term which is sometimes called an *Oseenlet*.

The resulting force is more difficult to calculate. Let us cite the following expression for the drag force, found by Goldstein [GOL 29]:

$$F = 6\pi \mu U_\infty R \left[1 + \frac{3}{8}\mathrm{Re} - \frac{19}{320}\mathrm{Re}^2 + \frac{71}{2560}\mathrm{Re}^3 - \frac{30179}{2150400}\mathrm{Re}^4 + \ldots \right] \quad [2.26]$$

and that given by Proudman and Pearson [PRO 57]:

$$F = 6\pi \mu U_\infty R \left[1 + \frac{3}{8}\mathrm{Re} + \frac{9}{40}\mathrm{Re}^2 \ln \mathrm{Re} + O\left(\mathrm{Re}^2\right) \right] \quad [2.27]$$

It is common to use *empirical formulas*, which are more accurate, to derive a correction coefficient to modify the Stokes relaxation time $\tau_{\mathrm{v,St}}$ involved in equation [2.23]:

$$\tau_{v,emp} = \tau_{v,St} / C_V \,.$$

Carlson and Hoglund [CAR 61], for example, give the drag coefficient C_V as a function of the Mach number and the

Reynolds number, built from the diameter of the spheres: $\mathrm{Re}_D = 2\mathrm{Re}$:

$$C_v = \frac{\left(1+0.15\,\mathrm{Re}^{0.687}\right)\left[1+\exp\left(0.427M^{-4.63}-3\,\mathrm{Re}_D{}^{-0.88}\right)\right]}{1+M\,\mathrm{Re}_D\left[3.82+1.28\exp\left(-1.25\,\mathrm{Re}_D\,M^{-1}\right)\right]} \qquad [2.28]$$

There are other empirical formulas which we can use [COR 56]. Brown and Lawler [BRO 03] compiled many of the results from the whole of the last century, and extract from those data an empirical formula for C_v which is commonly used today.

2.1.2.4. *Effect of acceleration of the particles and history term*

The *effects of acceleration of the particles* lead to a modification of the force. An additional term also appears, which is due to the history of the moving particle in relation to the carrier fluid.

If we take the Stokes' theorem as a starting point, and replace U_∞ with $(\mathbf{v}_g - \mathbf{v}_p)$, we obtain the following formula for condensed particles in a gas:

$$\mathbf{F} = 6\pi\mu U_\infty R\left(\mathbf{v}_g - \mathbf{v}_p\right) + \frac{2}{3}\pi R^3 \rho_{gs}\left(\frac{d_p\mathbf{v}_g}{dt} - \frac{d_p\mathbf{v}_p}{dt}\right) - \frac{4}{3}\pi R^3 \nabla p + 6R^2\sqrt{\pi\mu\rho_{gs}}\int_t^0 \left(\frac{d_p\mathbf{v}_g}{dt'} - \frac{d_p\mathbf{v}_p}{dt'}\right)\Big/\sqrt{t-t'}\,dt' + \frac{4}{3}\pi R^3 \rho_{gs}\,\mathbf{g} \qquad [2.29]$$

The right-hand side of equation [2.29] contains the terms which describe all of the forces acting on a spherical particle. The meanings of all these terms are recapped below:

– drag force: force of viscous friction due to the relative velocity of the particle and the fluid;

– added mass force: additional force needed to move the fluid around an accelerating particle;

– pressure gradient: acceleration of the fluid transferred to the particle;

– historical force or *Basset force*: interaction of the particle with its own wake;

– Archimedes force: effect of gravity.

Note that in this case, generally, except at the vicinity of the critical point, we have $\rho_{gs} << \rho_{ps}$ and the unsteady additive terms (the *Basset–Boussinesq forces*) are completely negligible [DOD 95].

We can also cite the models of turbulent transport of inertial particles, at the scale of which the fluid–particle coupling essentially works *via* the same physical mechanisms.

Let us mention wake instabilities, which are not discussed here, which can have a direct impact, for example, on the effective drag [JEN 04].

It is also worth mentioning a limitation of the model equation [2.29], which assumes outright that the particle is a single point (at the scale of the inhomogeneities of the flow) and does not take account of the rotation of the particle. For instance, we need to be able to define the particle/fluid slip velocity, which requires the velocity of the fluid to be identical all around the particle. In order to go further, there are various extensions that can be used. Firstly, there are 1st-order corrections (known as the Faxén correction, described in [MAX 83] cited above, but also in [GAT 83], which takes account of a 2nd-order asymptotic expansion depending on the particle size, so as to include the local inhomogeneity of the flow). Beyond these low-order effects, though, the integration of the true effects of finite scale remains a real challenge – particularly in the turbulent regime (see, for example, the experimental studies by

[MAC 14] or [QUR 07]), but also for the simple problem of particle sedimentation or bubble rising.

Numerical strategies which resolve the particles precisely are beginning to appear in the literature, such as [HOM 10] or [UHL 05].

One final aspect, which will not be discussed here, relates to the *two-way coupling* and *four way-coupling* which can arise when the particle concentration increases, so that not only is the fluid acting on the particles, but the particles themselves are altering the flow (*two-way coupling*), or interacting with one another (*four-way coupling*).[1]

2.2. Heat exchanges

In the Stokes' theorem, the convective terms are neglected. If we do the same thing to investigate heat transfer, we are led to determine the heat flow for a spherical particle at rest, of temperature T_0, dipped into an immobile fluid environment at temperature T_∞, at a significant distance from the particle (Figure 2.4). The only phenomenon which takes place in this scenario is heat conduction into the fluid. The equation governing the problem is reduce to $\nabla^2 T = 0$, or, in spherical coordinates, $\frac{\partial}{\partial r}\left(r^2 \frac{\partial T}{\partial r}\right) = 0$.

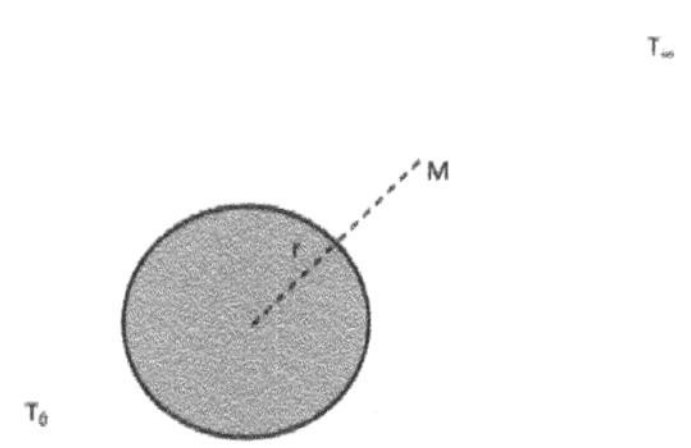

Figure 2.4. *Heat exchanges from a sphere*

1 The author wishes to thank Michael Bourgoin for his attentive reading and contributions to section 2.1.

The solution, for a sphere of radius of one unit, is:

$$T = T_\infty + \frac{T_0 - T_\infty}{r} \quad [2.30]$$

The heat flux through a sphere of any given radius r is constant:

$$Q = -4\pi r^2 \lambda \partial T/\partial r = 4\pi h (T_0 - T_\infty) \quad [2.31]$$

This corresponds to a flow of heat from the inside of the sphere toward the outside. The Nusselt number "Nu" is defined by:

$$\mathrm{Nu} = 2 R h/\lambda \quad [2.32]$$

where $2R$ is the diameter of the sphere, h the heat exchange coefficient per unit area at the surface of the sphere, and λ the heat conductivity. Thus, in the extreme case of the sphere at rest, we obtain: Nu =2.

In order to take into account the flow, dimensional analysis (see Appendix 1 of [PRU 13]) will give us:

$$\Pi_\kappa = \kappa/\nu, \Pi_h = h/(2R)^{(u-v)} U_\infty{}^v \nu^{(1-v)} \quad [2.33]$$

where $u = 2$ v-1 and, according to the Vashy-Buckingham theorem: $\Pi_h = \psi(\Pi_\kappa)$.

In order to take into account the fact that for $U_\infty = 0$, we obtain $\mathrm{Nu} = 2$, we need to take $(h - \lambda/R)$ instead of h to define Π_h. With a power law for ψ, we then obtain:

$$\mathrm{Nu} = 2 + A\,\mathrm{Re}^v\,\mathrm{Pr}^{(1-w)} \quad [2.34]$$

In a laminar regime, $\mathrm{Nu} = 1/2$, and for solid spheres, $1 - w = 1/3$.

The more elaborate formula proposed by Carlson and Hoglund, which takes into account the Mach number, gives us:

$$\mathrm{Nu} = 2\frac{1+0.230\,\mathrm{Re}^{0.55}}{1+6.84\,M\,\mathrm{Re}^{-1}\left(1+0.230\,\mathrm{Re}^{0.55}\right)} = 2C_T \qquad [2.35]$$

Clift *et al.* [CLI 78] give other formulas for Nu and also for the Sherwood number Sh, which is a Nusselt number for the diffusion of the species: $\mathrm{Sh} = 2R\,\alpha_C/\rho\,D$. So too do [ABR 89, LAW 77] in the presence of evaporation.

If we know the heat exchange coefficient per unit area, we can calculate the heat flux per unit volume of particles:

$$Q = \frac{3\alpha}{R}\left(T_g - T_p\right), \alpha = \frac{\kappa}{R}C_T \qquad [2.36]$$

In view of the phenomenological relation given later on for suspensions: $Q = \frac{\rho_{ps}\,c_c}{\tau_T}\left(T_g - T_p\right)$, we can therefore deduce the expression of the diffusion relaxation time: $\tau_T = R^2\rho_{ps}c_c/3\,\kappa\,C_T$

2.3. Combustion of a drop of fuel in an oxidizing environment

A variety of configurations may arise in laminar combustion of drops of pure fluid in a gaseous mixture.

To begin with, the droplet is experiencing evaporation in the presence of its neighbors at a sufficient distance so that there is no interaction between them. Combustion takes place quite far from the set of drops, and the thermal and chemical characteristics of the gaseous mixture produced by this combustion are known, constituting the conditions at infinity. This problem was dealt with in Appendix A4.3 of [PRU 12]. The aim there was to evaluate the role of the hypotheses made in irreversible thermodynamics extended over the unsteady process in the presence of an acoustic disturbance. A brief introduction to the quasi-steady regime was given on that occasion, assuming local equilibrium between evaporation and condensation at the interface.[2]

Remember the main results obtained with the *Shvab-Zeldovich approximation*, in the following cases: a drop of pure liquid in its own vapor [PRU 12], and a drop of pure liquid in a gaseous mixture with constant physical properties [PRU 10b, PRU 12]. In both cases, we obtain a "d^2" law for the regression of the diameter $d = 2\,r_s$ of the drop:

$$d^2 = d_0^{\,2} - K\,t \qquad [2.37]$$

with:

$$K = \frac{8}{\rho_L} \int_{T_s}^{T_\infty} \frac{\lambda(\theta)\,d\theta}{\int_{T_s}^{\theta} c_p(\theta)\,d\theta + l(T_s)}$$

in the case of the liquid's own vapor alone.

In the case of a gaseous mixture with constant physical properties:

$$K = \frac{8\lambda}{c_p} \ln\left[\frac{c_p\,(T_\infty - T_s)}{l} + 1\right]$$

The case of a spherical droplet closely surrounded by a flame was discussed in section 3.2.2.3 of [PRU 13] with the

2 The validity of this hypothesis is established for various propellants, in given conditions of temperature and pressure, in an article by Mauriot and Prud'homme [MAU 14], where the influence of the vapor recoil is also considered.

Shvab-Zeldovich approximations, in a quasi-steady regime. Remember that in that case:

$$K = \frac{8g}{\rho_L} \ln\left[\frac{(\beta_{O\infty} + \beta_{Ts})\Delta H}{l} + 1\right], \ \beta_j = Y_j / \nu_j \mathcal{M}_j, \ \beta_T = \sum_j Y_j \int_{T^0}^{T} c_{p,j}\, dT \Big/ \Delta H$$

with:

$$g = \lambda / c_{pf} = \rho D, \ \Delta H = -\nu_j \mathcal{M}_j \left(q_f^0\right)_j$$

The energy mobilized by the reaction with the flame is diffused toward the surface of the drop, thus causing its evaporation and the release of gaseous fuel F. The pressure is taken to be constant, and the temperature of the drop to be uniform and constant – this is the saturating vapor temperature T_l. At infinity, the temperature is T_∞ and the mass fraction of the oxidizing agent is $Y_{O\infty}$. The stoichiometric coefficients are taken to be equal to one for the purpose of simplification, but the theory remains the same even if different stoichiometric coefficients are used.

Remember that it is important to determine the evolution of the radius of the drop over time. Indeed, this yields knowledge of the combustion time and the length of the chamber. The ratio r_f / r_s of the radius of the flame (when the flame surrounds the drop) is also important. It indicates the possible interactions between the combusting drops in the flow. The above theories are, in fact, applicable only in the absence of interaction.

The effects of coupling between the flow and the individual process of the combusting drop, though, are much more difficult to evaluate.

Other cases are dealt with in works on combustion [WIL 85, PET 00, POI 05, PRU 10a].

3

Simplified Model of a Non-reactive Flow with Particles

For our discussion here, we shall assume that the dispersed phase is dilute and made up of only one type of particles – non-deformable and isothermal, *with no phase change or adsorption/desorption reactions*. The fluid is an inert gas similar to a perfect fluid, except in the case of extreme vicinity of the particles, upon which it exerts a friction force $\vec{\mathbf{F}}$ and to which it transfers the quantity of heat Q per unit time and in the unitary volume.

Aside from these exchange terms, each component behaves like a fluid: gas and fluid of particles, with respective velocities $\mathbf{v}_g$ and $\mathbf{v}_p$. The densities are reported to the volume of the mixture. The particles do not interact with one another, and therefore there is no inter-particle pressure.[1] In addition, we neglect distant forces such as gravity.

1 A classic example is that of a dusty gas, studied by Marble [MAR 70].

3.1. Variables characterizing the flow

A "particle" of mixture is a composite system at the macroscopic scale.[2] It has two components: the fluid phase and the condensed phase – a set of very small particles, similar to a fluid. In this section, we envisage a gas for the fluid phase, and the values relating thereto will be denoted by the subscript g, but the general balance equations established here will also be valid in the case of a liquid provided the law of state is not specified. The dispersed phase will be denoted by the subscript p and the condensed matter (solid or liquid, non-deformable) of which it is made up is represented by the subscript c. To begin with, we shall define specific values for each "fluid". For the particles (see Figure 3.1):

$$\rho_{ps} = \frac{m_p}{\mathcal{V}_p} = \frac{\mu_p}{4/3\pi R^3} \quad [3.1]$$

μ_p being the mass of a condensed particle and v_p its volume, equal to $(4/3\pi R^3)$ if it is spherical, with radius R. The quantity m_p is the total mass of the condensed species in a small control volume (the particle of the mixture) $\mathcal{V}_p$ of the condensed species.

Similarly, for the fluid phase, we shall have:

$$\rho_{gs} = \frac{m_g}{\mathcal{V}_g} \quad [3.2]$$

Let $\mathcal{V} = \mathcal{V}_g + \mathcal{V}_p$ be the volume of the mixture in question. We define the partial densities in the mixture:

---

2 This must not be confused with the particles of the multiphase medium, which may, *a priori*, be bubbles in a liquid, drops or dust grains in a gas or in a liquid.

$$\rho_p = \frac{m_p}{\mathcal{V}} = \varepsilon_p \rho_{ps} \quad [3.3]$$

$$\rho_g = \frac{m_g}{V} = \varepsilon_g \rho_{gs} \quad [3.4]$$

and the average density of the mixture: $\rho = \rho_p + \rho_g$.

The porosity of the mixture is:

$$\varepsilon = \varepsilon_g = \mathcal{V}_g / \mathcal{V} \quad [3.5]$$

We also have $\varepsilon_p = \mathcal{V}_p / \mathcal{V} = 1 - \varepsilon$, and in our subsequent discussion we shall accept the hypothesis that ε is near to 1.

On average, in a macroscopic particle of the mixture, the condensed phase will have the velocity $\mathbf{v}_p$ and the gaseous phase the velocity $\mathbf{v}_g$.

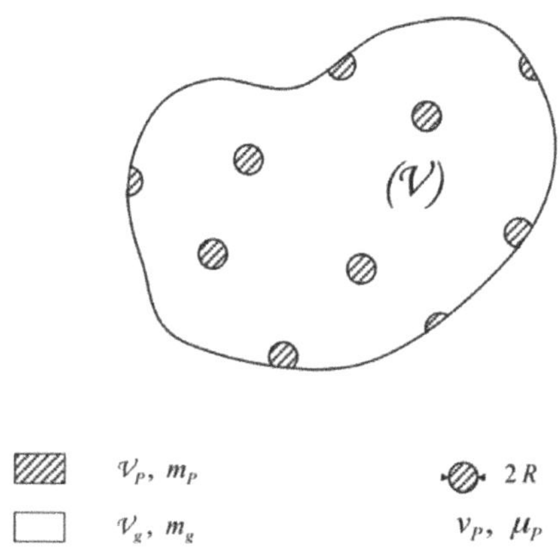

Figure 3.1. *A control volume of the two-phase mixture*

The thermodynamic state of the *condensed phase* will be defined solely by the temperature T_p, whereas that of the fluid phase will be defined by T_g and ρ_{gs}. Thus, for the internal energy values per unit mass of the dispersed phase, we have:

$$e_{ps} = e_{ps}\left(T_p\right), \quad de_{ps} = T_p\, ds_{ps} = c_c\, dT_p \qquad [3.6]$$

where c_c is the mass specific heat of the condensed phase.

For the gaseous phase:

$$e_{gs} = e_{gs}\left(T_g, \rho_g\right), \quad de_{gs} = T_g\, ds_{gs} - p\, d\vartheta_g \qquad [3.7]$$

where $\vartheta_g = 1/\rho_g$ and if we accept that $\vartheta_{gs} \cong \vartheta_g$ because of the supposedly high degree of porosity: $\varepsilon \cong 1$.

If the fluid phase is a perfect gas, the fundamental energy law is deduced from the following laws of state:

$$p = \rho_g r_g T_g, \quad de_g = c_v\, dT_g \qquad [3.8]$$

If the condensed particles are non-deformable, two types of exchanges take place between the condensed phase and the fluid phase: the force of friction $\vec{\mathbf{F}}$ and quantity of heat Q are passed from the fluid to the particles per unit volume of the mixture. As each component has its own velocity, we can define the material derivatives:

$$d_p/dt = \partial/\partial t + \mathbf{v}_p \cdot \nabla, \quad d_g/dt = \partial/\partial t + \mathbf{v}_g \cdot \nabla \qquad [3.9]$$

3.2. Balance equations

As the carrier fluid (g) and the dispersed phase (p) have different motions, it is necessary to establish balance equations for each phase.

3.2.1. *Balances for the particles*

For the fluid of particles, in the absence of interaction between the condensed particles, the inter-particle pressure plays no part. The temperature is, on average, that of the condensed particles, within which it is supposed to be uniform.

Thus, by tracking the movement of a connected domain $\mathcal{V}_p$ of the dispersed phase (the particulate fluid), in the absence of mass exchanges with the carrying fluid (no evaporation or condensation in this highly simplified model[3]), we have:

$$\frac{d_p}{dt}\int_{\mathcal{V}_p} \rho_p \, d\mathcal{V} = 0 \qquad [3.10]$$

from which we can deduce the mass balance, as we did in Chapter 2 of [PRU 12]:

$$\frac{d_p \rho_p}{dt} + \rho_p \nabla \cdot \mathbf{v}_p = 0 \qquad [3.11]$$

The integral momentum balance, in accordance with our simplified hypotheses (absence of distant forces and of the pressure gradient on the condensed phase), in view of the friction force exerted on the unit volume of the particulate fluid $\mathcal{F} = \rho_p \mathbf{F} / \mu_p$, will be:

$$\frac{d_p}{dt}\int_{\mathcal{V}_p} \rho_p \mathbf{v}_p \, d\mathcal{V} = \int_{\mathcal{V}_p} \mathcal{F} \, d\mathcal{V} \qquad [3.12]$$

3 For phase changes, see Chapter 4.

Bearing in mind the mass balance, we deduce from this the local momentum balance[4]:

$$\rho_p \frac{d_p \mathbf{v}_p}{dt} = \mathcal{F} \qquad [3.13]$$

For the energy, we need to take account of the heat flux $\mathcal{Q} = \rho_p Q / \mu_p$ passed on by the gas to the condensed species, and of the power of the force $\mathcal{F}$:

$$\frac{d_p}{dt} \int_{\mathcal{V}_p} \rho_p \left(e_{ps} + \mathbf{v}_p^{\,2}/2 \right) d\mathcal{V} = \int_{\mathcal{V}_p} \left(\mathcal{Q} + \mathcal{F} \cdot \mathbf{v}_p \right) d\mathcal{V} \qquad [3.14]$$

From this, we deduce the local energy equation:

$$\rho_p \frac{d_p \left(e_{ps} + \mathbf{v}_p^{\,2}/2 \right)}{dt} = \mathcal{Q} + \mathcal{F} \cdot \mathbf{v}_p \qquad [3.15]$$

With scalar multiplication by $\mathbf{v}_p$ of both sides of the momentum equation and subtracting the result, term by term, from the energy equation, we obtain the local internal energy equation:

$$\rho_p \frac{d_p e_{ps}}{dt} = \mathcal{Q} \qquad [3.16]$$

Note that $\mathcal{F}$ and $\mathcal{Q}$, which are generalized flows – in the sense of TIP – of gas and condensed species, must in fact be considered as production factors for the mixture.

4 In the presence of gravity and the effect of the pressure gradient on the condensed phase, we would have: $\rho_p \frac{d_p \mathbf{v}_p}{dt} = \mathcal{F} - \alpha_p \nabla p + \rho_p \mathbf{g}$, where α_p is the phase fraction, instead of equation [3.13]. Equations [3.14] to [3.16] will therefore need to be modified.

3.2.2. *Balances for the gaseous phase*

In order to establish the balances for the gas, this time we need to track the motion of a domain ($\mathcal{V}_g$).

The equations obtained are thus locally and successively, for the mass, momentum, energy and internal energy:

$$\begin{cases} \dfrac{d_g \rho_g}{dt} + \rho_g \nabla \cdot \mathbf{v}_g = 0 \\ \rho_g \dfrac{d_g \mathbf{v}_g}{dt} + \nabla p = -\mathcal{F} \\ \rho_g \dfrac{d_p \left(e_{ps} + \mathbf{v}_g{}^2/2 \right)}{dt} + \nabla \cdot \left(p \mathbf{v}_g \right) = -Q - \mathcal{F} \cdot \mathbf{v}_g \\ \rho_g \dfrac{d_g e_{gs}}{dt} + p \nabla \cdot \mathbf{v}_g = -Q - \mathcal{F} \cdot \left(\mathbf{v}_p - \mathbf{v}_g \right) \end{cases} \quad [3.17]$$

because the heat flows are equal in modulus and have opposite signs to those given in section 3.2.1, and the point of application of the force shifts along with the condensed phase.[5]

3.2.3. *Entropy balance and phenomenological relations*

The expression of the entropy production rate of the mixture is important for the determination of the additional laws which give us the expressions of $\mathcal{F}$ and Q.

---

5 In the presence of gravity and the effect of the pressure gradient on the condensed phase, we would have: $\rho_g \dfrac{d_g \mathbf{v}_g}{dt} = \mathcal{F} - \alpha_g \nabla p + \rho_g \mathbf{g}$, where α_g is the gaseous phase fraction. The energy equations concerning the gaseous phase will therefore need to be modified.

Neither the particles nor the gas are home to irreversible processes, so the only irreversible processes take place in the gas/particle exchanges. Thus, we have:

$$\rho_p \frac{d_p s_{ps}}{dt} + \rho_g \frac{d_g s_{gs}}{dt} = \dot{W}_s \geq 0 \qquad [3.18]$$

The thermodynamic laws, with $\rho_{gs} \cong \rho_g$, lead to:

$$\begin{cases} d\, s_{ps} = T_p^{-1}\, de_{ps},\, de_{ps} = c_c\, dT_p \\ d\, s_{gs} = T_g^{-1}\, de_{gs} + pT_g^{-1}\, d\left(1/\rho_g\right), de_{gs} = c_v\, dT_g \end{cases} \qquad [3.19]$$

Using these Gibbs relations and the balance equations, we thus obtain:

$$\dot{W}_s = \left(\frac{1}{T_p} - \frac{1}{T_g}\right) Q + \frac{1}{T_g} \mathcal{F} \cdot \left(\mathbf{v}_g - \mathbf{v}_p\right) \geq 0 \qquad [3.20]$$

The principles of TIP enable us to envisage the following linear laws when the generalized forces and the generalized fluxes are not too significant:

$$Q = L_1 \left(\frac{1}{T_p} - \frac{1}{T_g}\right), \mathcal{F} = L_2 \frac{1}{T_g}\left(\mathbf{v}_g - \mathbf{v}_p\right) \qquad [3.21]$$

where L_1, L_2 are the phenomenological coefficients.

The internal energy and momentum balances of the dispersed phase therefore give us:

$$\begin{cases} \rho_p c_c d_p T_p / dt = \left(L_1 / T_g T_p\right) \left(T_g - T_p\right) \\ \rho_p d_p \mathbf{v}_p / dt = \left(L_2 / T_g\right) \left(\mathbf{v}_g - \mathbf{v}_p\right) \end{cases} \qquad [3.22]$$

Using these relations, we can define, respectively, the thermal relaxation time: $\tau_T = \rho_p c_c T_g T_p / L_1$ and the friction relaxation time: $\tau_v = \rho_g T_g / L_2$. Note that the τ_T / τ_v ratio is similar to a Prandtl number. The balance equations can therefore be written in their final form, which gives us the following system[6]:

$$\begin{cases} \dfrac{d_p \rho_p}{dt} + \rho_p \nabla \cdot \mathbf{v}_p = 0, \quad \dfrac{d_g \rho_g}{dt} + \rho_g \nabla \cdot \mathbf{v}_g = 0 \\ \rho_p \dfrac{d_p \mathbf{v}_p}{dt} + \rho_g \dfrac{d_g \mathbf{v}_g}{dt} + \nabla p = \mathbf{0} \\ \rho_p \dfrac{d_p \left(e_{ps} + |\mathbf{v}_p|^2 / 2 \right)}{dt} + \rho_g \dfrac{d_p \left(e_{gs} + |\mathbf{v}_g|^2 / 2 \right)}{dt} + \nabla \cdot \left(p \mathbf{v}_g \right) = 0 \\ \dfrac{d_p \mathbf{v}_p}{dt} = \dfrac{\mathbf{v}_g - \mathbf{v}_p}{\tau_v}, \quad \dfrac{d_p T_p}{dt} = \dfrac{T_g - T_p}{\tau_T} \end{cases} \quad [3.23]$$

With regard to the entropy production rate, it becomes:

$$\dot{W}_s = \frac{\rho_p c_c}{\tau_T T_p T_g} \left(T_g - T_p \right)^2 + \frac{\rho_p}{\tau_v T_g} \left(\vec{\mathbf{v}}_g - \vec{\mathbf{v}}_p \right)^2 \geq 0 \quad [3.24]$$

Two limited cases need to be considered.

– *At equilibrium* between the gas and the particles, the generalized forces vanish, which means $T_g = T_p$ and $\mathbf{v}_g = \mathbf{v}_p$; the particles, which are very small, are carried along by the gas and have the same temperature as it. The balance equations for the mixture are identical to those of a perfect fluid of velocity $\mathbf{v} = \mathbf{v}_g = \mathbf{v}_p$, density $\rho = \rho_g + \rho_p$, pressure p and

6 The second and third equations will need to be modified if we wish to take account of the weight of the mixture and the power associated therewith.

temperature $T = T_g = T_p$. The laws of state obtained are different to those of the gas.

From the mass balance equations for the particles and the gas, we can deduce (because the velocity vectors are the same): $d\rho_p / \rho_p = d\rho_g / \rho_g$, for each particle of the mixture. It follows from this that $\rho_p = \kappa \rho_g$.

If the coefficient κ is the same for all the particles and the gas is perfect, we obtain:

$$p = \rho \frac{r_g}{1+\kappa} T = \rho \bar{r} T, \quad e = \frac{e_g + \kappa e_p}{1+\kappa} = \bar{c}_v T \qquad [13.25]$$

These results show that in this case, the mixture behaves like a perfect gas, with specific heat $\bar{c}_v = (c_v + \kappa c_c)/(1+\kappa)$ and molar mass $\bar{\mathcal{M}} = (1+\kappa)\mathcal{M}_g$.

– *Another situation* where $\dot{W}_S = 0$ is where the generalized fluxes Q and $\mathcal{F}$ are null, which, in light of equations [3.23] and [3.24], gives us $T_p = Const.$ and $\mathbf{v}_p = Const.$ The motion of the particles is not influenced by that of the gas, and the exchanges between the two are more or less fixed. Such is the case with very large particles.

Apart from these two extreme cases, it is necessary to take into account all the equations governing the problem, and the relaxation times τ_T and τ_v come into play.

3.3. Application to the linearized study of sound propagation in a non-reactive dilute suspension

The linearized study of sound propagation was presented in [PRU 12] for an inert gas and in [PRU 13] for reactive gaseous mixtures. Here, we shall examine a suspension of non-reactive particles in an inert surrounding gas. Readers will note various points in common with the case of the homogeneous monoreactive gaseous mixture.

The reference configuration is that of equilibrium between the gas and the particles. Thus, it is characterized by: $\mathbf{v}_{g0} = \mathbf{v}_{p0} = 0$, T_0, $\rho_{p0} = \kappa \rho_{g0}$, $p_0 = \rho_{g0} r_g T_0$ (we suppose the gas to be perfect). The disturbances are denoted by the subscript (1), which gives us the following (see [KUE 73]):

$$v_g = v_{g1},\ \ v_p = v_{p1},\ \ T_g = T_0 + T_{g1},\ \ T_p = T_0 + T_{p1},\ \ p = p_0 + p_1,$$
$$\rho_g = \rho_{g0} + \rho_{g1},\ \ \rho_p = \kappa \rho_{g0} + \rho_{p1}.$$

The system describing the small motions is therefore as follows [KUE 73]:

$$\begin{cases} a)\ \partial \rho_{g1}/\partial t + \rho_{g0} \nabla \cdot \mathbf{v}_{g1} = 0, \qquad b)\ \partial \rho_{p1}/\partial t + \rho_{p0} \nabla \cdot \mathbf{v}_{p1} = 0, \\ c)\ \rho_{g0}\, \partial \mathbf{v}_{g1}/\partial t + \kappa \rho_{g0}\, \partial \mathbf{v}_{p1}/\partial t + \nabla p_1 = \mathbf{0}, \\ d)\ \rho_{g0}\, \partial e_{g1}/\partial t + \kappa \rho_{g0}\, \partial e_{p1}/\partial t + p_0 \nabla \cdot \mathbf{v}_{g1} = 0, \\ e)\ \partial \mathbf{v}_{p1}/\partial t = (\mathbf{v}_{g1} - \mathbf{v}_{p1})/\tau_v, \qquad f)\ \partial T_{p1}/\partial t = (T_{g1} - T_{p1})/\tau_T, \\ g)\ p_1 = r_g T_0 \rho_{g1} + r_g \rho_{g0} T_{g1}, \qquad h)\ e_{gs1} = c_v T_{g1} = (c_p/\gamma) T_{g1}, \\ i)\ e_{ps1} = c_c T_{p1} = \beta c_p T_{p1}, \quad \beta = c_c/c_p \end{cases} \qquad [3.26]$$

Equation [3.26b] is the only equation involving ρ_{p1}. Equation [3.26d] can be rewritten by taking into account [3.26h] and [3.26i]:

$$\partial T_{g1}/\partial t + \kappa\gamma\beta\, \partial T_{p1}/\partial t + (\gamma - 1) T_0 \nabla \cdot \mathbf{v}_{g1} = 0 \qquad [3.27]$$

By combining [3.27] with [3.26a], we obtain the first integral:

$$T_{g1} + \kappa\gamma\beta T_{p1} - (\gamma - 1) T_0\, \rho_{g1}/\rho_{g0} = 0 \qquad [3.28]$$

and by combining this final equation with [3.26g], we find:

$$p_1 = \gamma r_g T_0 \rho_{g1} - r_g \rho_{g0} \kappa\gamma\beta T_{p1} \qquad [3.29]$$

The entropy production rate, given in section 3.2, is null to the second order. It follows that: $\kappa \partial s_{p1}/\partial t + \partial s_{g1}/\partial t = 0$, or indeed: $\kappa s_{p1} + s_{g1} = 0$, and because $s_g = c_v \ln T_g - r_g \ln \rho_g + Const.$ and $s_p = c_c \ln T_{pg} + Const.$, we therefore find: $\kappa s_{p1} + s_{g1} = \kappa\beta c_p T_{p1}/T_0 + c_v T_{g1}/T_0 - r_g \rho_{g1}/\rho_{g0} = 0$, which simply yields the same first integral as equation [3.28], but gives us nothing new.

Equations [3.26f] and [3.28] give us:

$$\frac{\partial T_{p1}}{\partial t} = \frac{1}{\tau_T}\left[\frac{\gamma-1}{\rho_{g0}} T_0 \rho_{g1} - (1+\kappa\gamma\beta) T_{p1}\right] \qquad [3.30]$$

The system needing to be solved therefore is:

$$\begin{cases} \partial \rho_{g1}/\partial t + \rho_{g0} \nabla \cdot \mathbf{v}_{g1} = 0, \quad \partial \mathbf{v}_{p1}/\partial t = \left(\mathbf{v}_{g1} - \mathbf{v}_{p1}\right)/\tau_v, \\ \rho_{g0} \partial \mathbf{v}_{g1}/\partial t + \kappa \rho_{g0} \partial \mathbf{v}_{p1}/\partial t + \gamma r_g T_0 \nabla \rho_{g1} \\ -r_g \rho_{g0} \kappa\gamma\beta \nabla T_{p1} = \mathbf{0}, \\ \dfrac{\partial T_{p1}}{\partial t} = \dfrac{1}{\tau_T}\left[\dfrac{\gamma-1}{\rho_{g0}} T_0 \rho_{g1} - (1+\kappa\gamma\beta) T_{p1}\right] \end{cases} \qquad [3.31]$$

By eliminating ρ_{g1} and $\mathbf{v}_{g1}$, we are left with:

$$\begin{cases} \dfrac{\partial}{\partial t}\left[\tau_T \dfrac{\partial T_{p1}}{\partial t} + (1+\kappa\gamma\beta) T_{p1}\right] + (\gamma-1) T_0 \nabla \cdot \left(\tau_v \dfrac{\partial \mathbf{v}_{p1}}{\partial t} + \mathbf{v}_{p1}\right) = 0 \\ \dfrac{\partial}{\partial t}\left[\tau_v \dfrac{\partial \mathbf{v}_{p1}}{\partial t} + (1+\kappa)\mathbf{v}_{p1}\right] + \dfrac{\gamma r_g}{\gamma-1} \vec{\nabla}\left[\tau_T \dfrac{\partial T_{p1}}{\partial t} + (1+\kappa\beta) T_{p1}\right] = 0 \end{cases} \qquad [3.32]$$

Thus, we can set:

$$T_{p1} = \hat{T}_p \exp i(\omega t - \mathbf{K}\cdot\mathbf{x}), \quad \mathbf{v}_{p1} = \hat{\mathbf{v}}_p \exp i(\omega t - \mathbf{K}\cdot\mathbf{x})$$

The system [3.32] only has a solution if the following complex equation is satisfied:

$$c_0^2 \frac{K^2}{\omega^2} = \frac{(1+\kappa\gamma\beta + i\omega\tau_T)(1+\kappa+i\omega\tau_v)}{(1+\kappa\beta+i\omega\tau_T)(1+i\omega\tau_v)} \qquad [3.33]$$

with $c_0^2 = \gamma r_g T_0$ being the square of the celerity of sound in the gas.

The complex number K, which is the modulus of the complex wave number vector $\mathbf{K}$, is such that: $K/\omega = 1/c(\omega) + i\gamma(\omega)$, where c is the celerity of the wave and $\omega\gamma(\omega)$ its damping per unit length (similarly to section 1.1.2 of [PRU 13]).

When the frequency ω is close to zero, we obtain a real value:

$$\frac{c_0^2}{c_e^2} = \frac{(1+\kappa\gamma\beta)(1+\kappa)}{(1+\kappa\beta)} \qquad [3.34]$$

The disturbance is sufficiently slow for the mixture to have time to return to equilibrium. The celerity c takes its equilibrium value c_e.

As ω tends toward infinity, we also find a real value – this time equal to c_0. The disturbance is so quick that the particles do not have time to react. We are witnessing the case of freezing, and the celerity of sound is that of the gas.

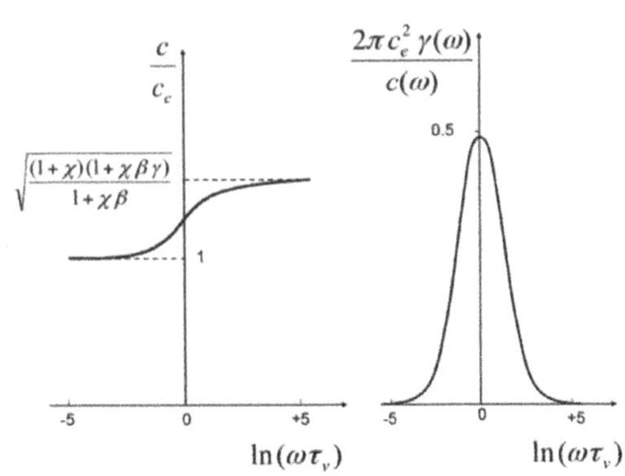

Figure 3.2. *Dispersion and absorption curves for a dilute suspension*

The shape of the curve $c(\omega)$ depends on the respective orders of magnitude of τ_T and τ_v. Note that $c_0 > c_e$ because $\gamma > 1$. For very small, light particles, the celerity of sound will, for finite frequencies, be close to c_e. For larger particles, however, which are more difficult to move, it will be nearer to c_0.

Figure 3.2 shows the results for particular values of the coefficients when the Prandtl number is equal to 2/3, which corresponds to $\tau_T = \beta\tau_v$.

3.4. Two-phase dilute flows in nozzles

The hypothesis of quasi-one-dimensional flow was used in Chapter 1 of [PRU 13]. We suppose that the parameters are approximately uniform in any cross-section $S(x)$. For the flows with particles in tubes with a constant cross-section, readers can usefully refer to [KLI 60] and, for plasmas, the next volume, *Flows and Chemical Reactions in an Electromagnetic Field* [PRU 14b] is instructive, as is the discussion of radiative media in Chapter 4 of this book.

In the steady regime, the general equations form the following system:

$$\begin{cases} \rho_p U_p S(x) = \dot{m}_p ,\ \rho_g U_g S(x) = \dot{m}_g \\ \rho_p U_p \dfrac{dU_p}{dx} + \rho_g U_g \dfrac{dU_g}{dx} + \dfrac{dp}{dx} = 0 \\ \rho_p U_p \dfrac{d\left(c_c T_p + U_p^{\ 2}/2\right)}{dx} + \rho_g U_g \dfrac{d\left(c_p T_g + U_g^{\ 2}/2\right)}{dx} = 0 \\ U_p \dfrac{dU_p}{dx} = \dfrac{U_g - U_p}{\tau_v} ,\ U_p \dfrac{dT_p}{dx} = \dfrac{T_g - T_p}{\tau_T} \end{cases} \qquad [3.35]$$

In this system, the first line corresponds to the conservation of flow rate of the particles and of the gas, the second to the momentum of the suspension, the third to its energy and the fourth to the velocity relaxation and thermal relaxation.

3.4.1. *Flow with constant phase shifts*

Generally speaking, we set $\dot{m}_p = \kappa \dot{m}_g$. The quantities $\dot{m}_g$ and κ are constants.

Here, we shall not discuss the general solution which results from numerical integration. We have a problem similar to that of the reactive flow discussed in Chapter 1 of [PRU 13], this time with two relaxation phenomena. As a matter of preference, here, we shall seek analytical solutions without restricting ourselves to the profile of a nozzle $S(x)$ – this will be a consequence of the hypotheses.

A well-known example is that of a nozzle with constant phase shifts [KLI 60]. We suppose that:

$$n = U_p / U_g = Const., \qquad q = \left(T_0 - T_p\right) / \left(T_0 - T_g\right) = Const. \qquad [3.36]$$

The velocity relaxation equation thus gives us:

$$U_p = (1-n)x/n\tau_v, \quad U_g = (1-n)x/n^2\tau_v \qquad [3.37]$$

The thermal relaxation equation is written thus, accepting that at $x = 0$, $U_p = U_g = 0$ and $T_p = T_g = T_0$: $U_p\, dT_p/dx = (q-1)(T_0 - T_p)/q\tau_T$.

However, the energy equation is integrated to give the following, in light of the expressions of U_p and U_g[7]:

$$T_0 - T_p = \frac{(1+\kappa n^2)q}{2(c_p + \kappa c_c q)}\left(\frac{1-n}{n^2\tau_v}\right)^2 x^2 \qquad [3.38]$$

By replacing $(T_0 - T_p)$ with this expression of the form $T_0 a^2 x^2$ and U_p with the value found above, in the thermal relaxation equation, we obtain the following condition between the phase shifts q and n:

$$2(1-n)/n\tau_v + (q-1)/q\tau_T = 0 \qquad [3.39]$$

The temperature profile T_g is therefore such that:

$$T_0 - T_g = \frac{1+\kappa n^2}{2(c_p + \kappa c_c q)}\left(\frac{1-n}{n^2\tau_v}\right)^2 x^2 \qquad [3.40]$$

Using the equation of state of the gas and the momentum equation, we can determine the profile of the nozzle, the

7 N.B. In this equation, c_p is the specific heat of the gas (not of the particles), at a constant pressure per unit mass, and c_c is the specific heat of the condensed species per unit mass.

evolution of the pressure and of the densities ρ_p and ρ_g. The calculation requires simple squaring to find *S(x)*. Thus, the momentum equation is written:

$$S\,dp/dx + \dot{m}_g(1-n)(1+\kappa n)/n^2\tau_v = 0$$

The partial density of the gas is: $\rho_g = \dot{m}_g/SU_g = \dot{m}_g n^2 \tau_v/S(1-n)x$ and the equation of state gives us:

$$p = \rho_g r_g T_g = \dot{m}_g n^2 \tau_v r_g T_0\left(1-a^2x^2\right)/S(1-n)x$$

Hence, dp/dx and the differential equation:

$$\left(1-a^2x^2\right)x\,dS/dx + \left(1-b^2x^2\right)S = 0 \qquad [3.41]$$

where:

$$b^2 = \frac{1}{T_0}\left(\frac{1-n}{n^2\tau_v}\right)^2 \frac{1+\kappa n}{r_g}\left(1-\frac{1+\kappa n^2}{1+\kappa n}\frac{r_g}{2\left(c_p+\kappa c_c q\right)}\right) \qquad [3.42]$$

We shall assume this quantity to be positive, and:

$$a^2 = \frac{1}{T_0}\left(\frac{1-n}{n^2\tau_v}\right)^2 \frac{1+\kappa n^2}{2\left(c_p+\kappa c_c q\right)}$$

The throat of the nozzle corresponds to the abscissa value $x_c = 1/b$.

By setting:

$$\alpha = \frac{1}{2}\left(\frac{b^2}{a^2}-1\right) = \frac{c_p+\kappa c_c q}{r_g}\frac{1+\kappa n}{1+\kappa n^2} - 1$$

(we suppose $a > 0$), we obtain:

$$\begin{cases} S(x) = C\left(1 - a^2 x^2\right)^{-\alpha} x^{-1} \\ dS/dx = C\left(1 - a^2 x^2\right)^{-(\alpha+1)} x^{-2}\left(b^2 x^2 - 1\right) \end{cases} \quad [3.43]$$

The profile of such a nozzle is shown in Figure 3.3; the convergent and divergent parts are very open, and the throat is fairly long.

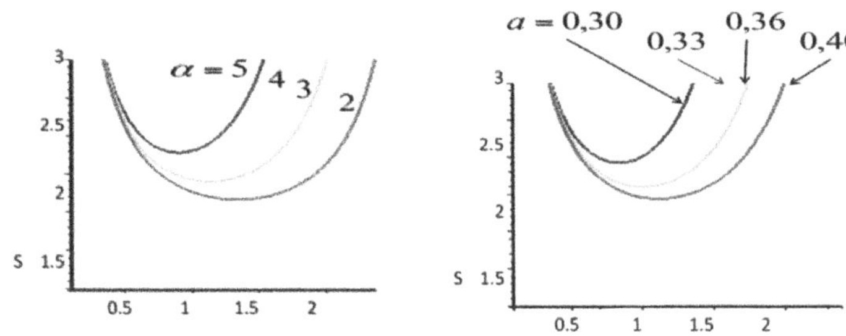

Figure 3.3. *Profile of a nozzle with constant phase shifts: left – influence of α for $a = 1/3$; right – influence of a for $\alpha = 4$*

3.4.2. *Numerical solutions*

Numerous authors [LYP 61, GLA 61, HUL 65] have performed calculations for two-phase flows in nozzles with particular profiles.

In real nozzle flows, the phase shifts are not constant, as shown by Figure 3.4, where, notably, we can clearly see that n is less than 1.

To suppose that $n < 1$ is a justified hypothesis in accelerated flows. This ensures that the right-hand side of equation [3.42] is indeed positive, as is the coefficient α.

The presence of a singular point in the vicinity of the throat, but upstream of it, is characteristic of the relaxing flows (see the case of reactive flows in Chapter 1 of [PRU 13]). This singularity corresponds to the equality between the velocity of the flow and the frozen celerity of sound. The fact that this does not occur exactly at the point of the throat makes it harder to determine the critical flow rates.

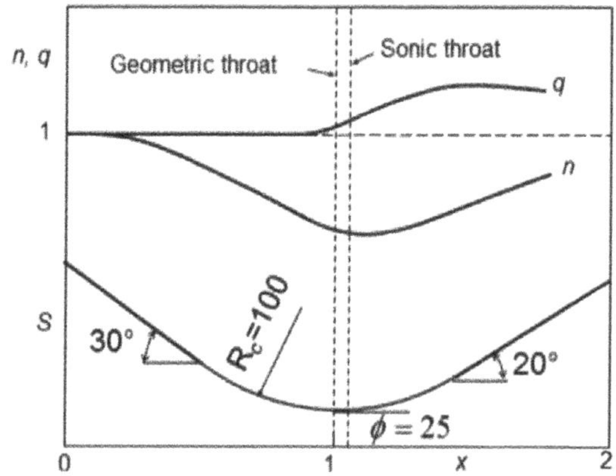

Figure 3.4. *Phase shifts in a nozzle, according to [KUE 73]:*

$$\rho_p \frac{d_p \mathbf{v}_p}{dt} = \mathcal{F} - \alpha_p \nabla p + \rho_p \mathbf{g}$$

4

Simplified Model of a Reactive Flow with Particles

For our discussion in this chapter, the particles are drops of fuel (e.g. a hydrocarbon) undergoing evaporation; the vapor phase is likely to react chemically with a gaseous oxidizing agent. The opposite situation arises, however, in cryogenic engines, where there are drops of oxidizing agent (liquid oxygen), whose vapor reacts with the fuel (gaseous hydrogen).

We shall first establish the system of balance equations for a spray of evaporating droplets by the method used previously for neutral particles.[1] Then we shall give those used in a conventional numerical code, and their possible extension to the case of drops with an internal temperature field. Then we shall apply the theory to the case of a fog flame.

4.1. Balance equations for a reactive fog

The two-phase mixture envisaged here is a suspension of droplets of pure liquid fuel in the process of evaporation in a

1 See Young's article [YOU 95] for the intermediary case of flows of gas and evaporating drops, but without chemical reactions.

reactive gaseous mixture comprising: the fuel vapor, an oxidizing agent and combustion products.

There have been many publications devoted to the modeling of these fogs, because of their importance in rocket propulsion and internal combustion engines [BOR 95, DEL 99, FRA 07, WIL 85].

Here, we shall return to using the approach developed in [PRU 10a] and [PRU 96], which consists of first writing the balance equations for the droplets and for the suspension, before deducing the equations for the gaseous mixture. The definitions given in section 3.1 remain valid: it is assumed that the density of the gaseous mixture is much lower than that of the liquid in the drops.

Another case would be that of particles with surface reactions, but this scenario shall not be discussed here.

4.1.1. *Balances for the droplets*

The number of particles $n(\mathbf{x},t)$ obeys the integral conservation equation:

$$\frac{d_p}{dt}\int_{\mathcal{V}_p} n\, d\mathcal{V} = 0 \qquad [4.1]$$

The material derivative d_p/dt means that at every point of the volume $\mathcal{V}_p$, we follow the motion of the droplets. We can deduce from this the local conservation equation:

$$\partial n/\partial t + \nabla\cdot\left(n\mathbf{v}_p\right) = 0 \qquad [4.2]$$

The local mass balance of the droplets is then:

$$d_p\rho_p/dt + \rho_p\nabla\cdot\mathbf{v}_p = \dot{W}_p \qquad [4.3]$$

with $\rho_p = nM$, M denoting the mass of the drop and, for the rate of production per unit volume: $\dot{W}_p = -n\dot{M}$, the product of the number of drops by the rate of evaporation of each drop.

For the momentum, we begin with the equation of each drop $M\, d_p \mathbf{v}_p / dt = \mathbf{F}$ and from this we deduce, for the unitary volume of fluid in the droplets:

$$\rho_p\, d_p \mathbf{v}_p / dt = \mathcal{F} \qquad [4.4]$$

The balance of the mass internal energy $\bar{e}_p$ involves the quantity of heat passed to the unit volume Q, the mass enthalpy $\bar{h}_p$ and the surface mass enthalpy h_S.

To establish the balances for the gaseous phase, we first write the equations of conservation of the mixture, and then subtract the equations of the droplets, term by term.

All of the balance equations for the droplets are recapped in Table 4.1.

Conservation of the number of droplets	$\partial n/\partial t + \nabla \cdot (n\mathbf{v}_p) = 0$
Mass balance equation	$d_p\rho_p/dt + \rho_p \nabla \cdot \mathbf{v}_p = \dot{W}_p$
Momentum equation	$\rho_p\, d_p\mathbf{v}_p/dt = \mathcal{F}$
Internal energy equation	$\rho_p \frac{d_p \bar{e}_p}{dt} = Q + \dot{W}_p (h_S - \bar{h}_p)$
Kinetic energy equation	$\rho_p \frac{d_p k_p}{dt} = \mathcal{F} \cdot \mathbf{v}_p$
Total energy equation	$\rho_p \frac{d_p(\bar{e}_p + k_p)}{dt} = Q + \dot{W}_p (h_S - \bar{h}_p) + \mathcal{F} \cdot \mathbf{v}_p$

Table 4.1. *Balance equations for the droplets. The number of droplets per unit volume is n; the subscript p which refers to the particles, in this case, denotes the droplets; $\dot{W}_p = -n\dot{M}$ is the opposite of the evaporation rate per unit volume of the liquid phase; h_S is the mass enthalpy of the liquid at the surface of the drop; $\bar{h}_p$ is the average enthalpy of the drop corresponding to its average internal energy $\bar{e}_p$*

4.1.2. *Balances of the mixture*

The total mass is conserved:

$$\frac{d_p}{dt}\int_{\mathcal{V}}\rho_p\, d\mathcal{V}+\frac{d_g}{dt}\int_{\mathcal{V}}\rho_g\, d\mathcal{V}=0 \qquad [4.5]$$

which gives us the local balance equation:

$$\frac{d_p\,\rho_p}{dt}+\rho_p\nabla\cdot\mathbf{v}_p+\frac{d_g\,\rho_g}{dt}+\rho_g\nabla\cdot\mathbf{v}_g=0 \qquad [4.6]$$

The momentum is conserved, so in the absence of distant forces we have:

$$\frac{d_p}{dt}\int_{\mathcal{V}}\rho_p\,\mathbf{v}_p\, d\mathcal{V}+\frac{d_g}{dt}\int_{\mathcal{V}}\rho_g\,\mathbf{v}_g\, d\mathcal{V}+\int_{\partial V}\vec{\vec{\mathbf{P}}}\cdot\mathbf{n}\; dS=\mathbf{0} \qquad [4.7]$$

which, locally, gives us:

$$\frac{\partial(\rho_p\,\mathbf{v}_p)}{\partial t}+\nabla.(\rho_p\mathbf{v}_p\otimes\mathbf{v}_p)+\frac{\partial(\rho_g\,\mathbf{v}_g)}{\partial t}+\nabla\cdot(\rho_g\mathbf{v}_g\otimes\mathbf{v}_g)+\nabla\cdot\vec{\vec{\mathbf{P}}}=\mathbf{0} \qquad [4.8]$$

or indeed, by using the mass balance equations for the particles and the total mass:

$$\rho_p\frac{d_p\mathbf{v}_p}{dt}+\rho_g\frac{d_g\mathbf{v}_g}{dt}+\nabla.\vec{\vec{\mathbf{P}}}=\dot{W}_p(\mathbf{v}_p-\mathbf{v}_g) \qquad [4.9]$$

With regard to the chemical species, we separate out everything relating to the vapor produced by the particles from everything relating to the other gaseous species:

$$\frac{d_p}{dt}\int_{\mathcal{V}}\rho_p\, d\mathcal{V}+\frac{d_v}{dt}\int_{\mathcal{V}}\rho_v\, d\mathcal{V}=\int_{\mathcal{V}}\dot{W}_v\, d\mathcal{V} \qquad [4.10]$$

where $\dot{W}_v$ is the reaction rate of the vapor (produced by the liquid droplets):

$$\frac{d_j}{dt}\int_{\mathcal{V}} \rho_j \, d\mathcal{V} = \int_{\mathcal{V}} \dot{W}_j \, d\mathcal{V}, \quad j \neq v \qquad [4.11]$$

The equations can be transformed to give us:

$$\begin{cases} \dfrac{\partial \rho_p}{\partial t} + \nabla.(\rho_p \mathbf{v}_p) + \dfrac{\partial \rho_v}{\partial t} + \nabla \cdot (\rho_v \mathbf{v}_v) = \dot{W}_v \\ \dfrac{\partial \rho_j}{\partial t} + \nabla \cdot (\rho_j \mathbf{v}_j) = \dot{W}_j, \quad j \neq v \end{cases} \qquad [4.12]$$

We can also bring into play the mass fractions Y_j in the gaseous phase and the diffusion flux, giving:

$$\begin{cases} \dfrac{\partial \rho_p}{\partial t} + \nabla \cdot (\rho_p \mathbf{v}_p) + \dfrac{\partial (\rho Y_v)}{\partial t} + \nabla \cdot (\rho Y_v \mathbf{v}_g + \mathcal{J}_{Dv}) = \dot{W}_v \\ \dfrac{\partial (\rho Y_j)}{\partial t} + \nabla.(\rho Y_j \mathbf{v} + \mathcal{J}_{Dj}) = \dot{W}_j, \quad j \neq v \end{cases} \qquad [4.13]$$

The total energy equation is written as:

$$\frac{d_p}{dt}\int_{\mathcal{V}} \rho_p (\bar{e}_p + k_p) d\mathcal{V} + \frac{d_g}{dt}\int_{\mathcal{V}} \rho_g (e_g + k_g) d\mathcal{V} + \int_{\partial \mathcal{V}} (\mathbf{q} + \mathbf{v_g} \cdot \vec{\vec{\mathbf{P}}}) \cdot \mathbf{n} \, dS = 0 \qquad [4.14]$$

where $\bar{e}_p$ is the mass energy of the liquid making up the particles, or indeed:

$$\begin{aligned} &\frac{\partial}{\partial t}\left(\rho_p (\bar{e}_p + k_p)\right) + \nabla \cdot \left(\rho_p (\bar{e}_p + k_p) \mathbf{v}_p\right) + \frac{\partial}{\partial t}\left(\rho_g (e_g + k_g)\right) \\ &+ \nabla \cdot \left(\rho_g (e_g + k_g) \mathbf{v}_g\right) + \nabla \cdot \left(\mathbf{q} + \mathbf{v}_g . \vec{\vec{\mathbf{P}}}\right) = 0 \end{aligned} \qquad [4.15]$$

and, by using the material derivatives and the mass balance equations:

$$\rho_p \frac{d_p(\bar{e}_p + k_p)}{dt} + \rho_g \frac{d_g(e_g + k_g)}{dt} + \nabla \cdot \left(\mathbf{q} + \mathbf{v}_g \cdot \vec{\vec{\mathbf{P}}}\right) \qquad [4.16]$$
$$= -\dot{W}_p\left(\bar{e}_p + k_p - e_g - k_g\right)$$

4.1.3. *Gaseous balances*

As has been said earlier on, we subtract the equations for the particles from those for the mixture, term by term. This gives us Table 4.2.

Equation of the droplet vapor	$\rho_g \frac{d_g Y_v}{dt} + \nabla \cdot \mathcal{J}_{Dv} = \dot{W}_v - (1 - Y_v)\dot{W}_p$
Diffusion equation for the other species	$\rho_g \frac{d_g Y_j}{dt} + \nabla \cdot \mathcal{J}_{Dj} = \dot{W}_j + Y_j \dot{W}_p$
Mass balance of the gas	$\frac{d\rho_g}{dt} + \rho_g \nabla \cdot \mathbf{v}_g = -\dot{W}_p$
Momentum equation of the gas	$\rho_g \frac{d_g \mathbf{v}_g}{dt} + \nabla \cdot \vec{\vec{\mathbf{P}}} = -\mathcal{F} - \dot{W}_p\left(\mathbf{v}_p - \mathbf{v}_g\right)$
Internal energy equation	$\rho_g\, de_g/dt + \nabla \cdot \mathbf{q} = -\nabla \otimes \mathbf{v}_g : \vec{\vec{\mathbf{P}}} - \dot{W}_p\big[h_S - p/\rho_{ps}$ $-\left(\mathbf{v}_p - \mathbf{v}_g\right)^2/2\big] - Q - \mathcal{F} \cdot \left(\mathbf{v}_p - \mathbf{v}_g\right)$
Total energy equation	$\rho_g d_g\left(e_g + k_g\right)/dt + \nabla \cdot \left(\mathbf{q} + \mathbf{v}_g \cdot \vec{\vec{\mathbf{P}}}\right) =$ $-\dot{W}_p\left(h_S - p/\rho_{ps} + k_p - e_g - k_g\right) - Q - \mathcal{F} \cdot \mathbf{v}_p$

Table 4.2. *Balance equations for the gaseous phase. The energy equations involve the term* p/ρ_{ps}*, which comes from the equality* $\bar{h}_p = \bar{e}_p + p/\rho_{ps}$*, valid for the specific average values for each drop*

4.1.4. *Entropy balance of the spray and phenomenological relations*

In contrast with equation [3.18], the entropy balance equation of the fog contains additional terms. This gives us:

$$\rho_p \frac{d_p s_p}{dt} + \rho_g \frac{d_g s_g}{dt} - \dot{W}_p \left(s_g - s_{ps} \right) + \nabla \cdot \frac{\mathbf{q} - \sum_j g_j \mathcal{J}_{Dj}}{T_g} = \dot{W}_S \geq 0 \quad [4.17]$$

By proceeding in the same way as in the case of inert particles, i.e. using the fundamental thermo dynamic relations of the droplets and the gaseous mixture, we finally obtain the expression of the entropy production rate:

$$\begin{aligned} \dot{W}_S = {} & Q\left(\frac{1}{T_p} - \frac{1}{T_g} \right) + \frac{\mathcal{F}}{T_g} \cdot \left(\mathbf{v}_g - \mathbf{v}_p \right) + \mathbf{q} \cdot \nabla \frac{1}{T_g} - \sum_j \mathcal{J}_{Dj} \cdot \nabla \frac{g_j}{T_g} \\ & - \frac{1}{T_g} \mathbf{\Pi} : \nabla \otimes \mathbf{v}_g - \sum_j \frac{g_j}{T_g} \dot{W}_j - \dot{W}_p \left[\frac{h_S - \bar{h}_p}{T_p} - \frac{1}{T_g} \left(h_S - \frac{p_g}{\rho_{ps}} - e_g \right.\right. \quad [4.18] \\ & \left.\left. + \frac{1}{2} \left| \mathbf{v}_p - \mathbf{v}_g \right|^2 - \frac{p_g}{\rho_g} - \sum_j g_j Y_j - g_v + T_g \left(s_g - \bar{s}_p \right) \right) \right] \end{aligned}$$

By overlooking the relative kinetic energy and the term in $1/\rho_{ps}$, for the part of the entropy production rate relating to the gas/particle exchanges, we obtain:

$$\dot{W}_S^{g-p} = \left(\frac{1}{T_p} - \frac{1}{T_g} \right) \left(Q + \dot{W}_p h_S \right) + \mathcal{F} \cdot \frac{\mathbf{v}_g - \mathbf{v}_p}{T_g} - \dot{W}_p \left(\frac{\bar{g}_p}{T_p} - \frac{g_v}{T_g} \right) \quad [4.19]$$

This expression involves three sorts of products of the generalized flux by the generalized forces corresponding respectively to the temperature jump, the velocity jump and the jump in chemical potential between the gas and the particles. Linear TIP then enables us, in the absence of coupling between these phenomena, to deduce the following constitutive equations:

$$\begin{cases} Q+\dot{W}_p h_S = L_1\left(\dfrac{1}{T_p}-\dfrac{1}{T_g}\right), \quad \boldsymbol{\mathcal{F}} = \dfrac{L_2}{T_g}\left(\mathbf{v}_g-\mathbf{v}_p\right), \\ \dot{W}_p = L_3\left(\dfrac{g_v}{T_g}-\dfrac{\bar{g}_p}{T_p}\right) \end{cases} \qquad [4.20]$$

where the phenomenological coefficients L_1, L_2, L_3 need to be determined [PRU 10].

4.1.5. *Equations of the two-phase CEDRE solver*

The Eulerian solver SPIREE (simulation of inert or reactive particles in flow from the Eulerian point of view) of the CEDRE computation platform (calculation of two-phase reactive flows for the energy aspects) [REF 11, SEM 09] from ONERA[2], describes the behavior of dilute two-phase suspensions. The system of equations is that of "sprays", in which the pressure terms are absent. For a given class of particles, characterized by their diameter[3] and their density ρ_L, which depends only on the temperature T_L,

---

2 ONERA: *Office national d'études et de recherches aérospatiales* – French National Office for Aerospace Research.

3 In the case of polydisperse suspensions, it is helpful to define statistical sizes of the drops. Mugele and Evans [MUG 51] give a very general formula on the basis of the distribution function $f(D)$ in diameter:

$$D_{pq} = \left[\int_0^\infty D^p f(D)dD \Big/ \int_0^\infty D^q f(D)dD\right]^{1/(p-q)}$$

(also see [BAB 02]). For a set of N drops of common diameter d_i, we can write:

$$D_{pq} = \left[\sum_{i=1}^{N} d_i^{\,p} \Big/ \sum_{i=1}^{N} d_i^{\,q}\right]^{1/(p-q)}$$

for the characteristic diameter [DOR 12]. The most commonly used diameters are the average diameter D_{10} and the Sauter mean diameter D_{32} – the ratio proportional to the sum of the volumes over the sum of the areas.

the system considered is written in conservative variables $\rho, \rho\mathbf{u}, \rho e_t, N$ in the following form [MUR 06]:

$$\begin{cases} \dfrac{\partial \rho}{\partial t} + \nabla \cdot (\rho \mathbf{u}) = -n\dot{M} \\ \dfrac{\partial (\rho \mathbf{u})}{\partial t} + \nabla \cdot (\rho \mathbf{u} \otimes \mathbf{u}) = n\mathbf{F} - n\dot{M}\mathbf{u} \\ \dfrac{\partial (\rho e_t)}{\partial t} + \nabla \cdot (\rho e_t \mathbf{u}) = n(\mathbf{F} \cdot \mathbf{u} + Q) - n\dot{M}\left(h_S + \dfrac{1}{2}|\mathbf{u}|^2 \right) \\ \dfrac{\partial n}{\partial t} + \nabla \cdot (n\mathbf{u}) = 0 \end{cases} \quad [4.21]$$

In this system, **u** is the common velocity of the particles and the gaseous mixture, and the total energy per unit mass is defined as the sum of the internal energy and the kinetic energy: $e_t = e + \frac{1}{2}|\mathbf{u}|^2$. Q is the heat flux injected into each particle by the gas; $\mathbf{F}$ is the force exerted by the gas on each particle; $\dot{M}$ is the rate of evaporation of each particle; h_S is the enthalpy per unit mass of the liquid species evaporating; and ρ_L is the density specific to each particle.

The law of state of the particles is: $e_L = e_L^0 + c_s(T_L)\left(T_L - T_L^0\right)$ and their volume fraction is $\alpha = n\frac{\pi}{6}D^3$. The density of the particles in relation to the total volume is: $\rho = \alpha \rho_L$.

4.1.6. *Modified equations to take account of an internal temperature gradient of the drops: multi-layer model*

In the above model, each drop is characterized by its temperature T_L. This requires infinite heat conductivity. In reality, the temperature within each drop evolves in space and over time. If we accept the hypothesis of spherical symmetry, then $T_L = T_L(r,t)$. This hypothesis changes the response of the drops when subjected to stresses – e.g. acoustic disturbances (see [PRU 10b]).

A multi-layer model has been developed to describe these drops [PRU 10b]. On a thermal level, the continuous distribution of temperatures throughout the drop is replaced by a discrete distribution.

4.1.6.1. *Case of the 2-layer model*

The principle of the 2-layer model [PRU 03, PRU 04] consists of replacing the drop with the set of two uniform concentric sub-systems, as shown in Figure 4.1. The outer layer is at the temperature T_S and the central nucleus is at T_0. This outer layer receives a certain quantity of heat from the surrounding gas. Some of that received heat serves for the evaporation, and another part Q serves to heat the liquid, and it cedes the quantity of heat Q_{S0} to the central nucleus. The problem is how best to represent these heat exchanges so that the result of the calculation is as close as possible to the actual continuous evolution.[4]

An additional variable T_0 and a parameter ε are therefore added to the case of the infinite conductivity in the drops.

Thus, with $\alpha = N\pi D^3/6, \varepsilon = D_0^3/D^3$ and $\alpha_0 = \varepsilon\alpha,\ \alpha_S = (1-\varepsilon)\alpha$, we write:

$$\rho = \alpha_0\,\rho_L(T_0) + \alpha_S\,\rho_L(T) \qquad [4.22]$$

4 The exchange coefficients were determined by comparison with the continuous evolution, which can be resolved analytically in the case of the fed drop [PRU 10b].

where $\rho_L(T)$ is the liquid's specific density – a given function of the temperature, T the temperature of the external layer (called T_S in our models), in contact with the gaseous mixture and T_0 the temperature of the inner nucleus.

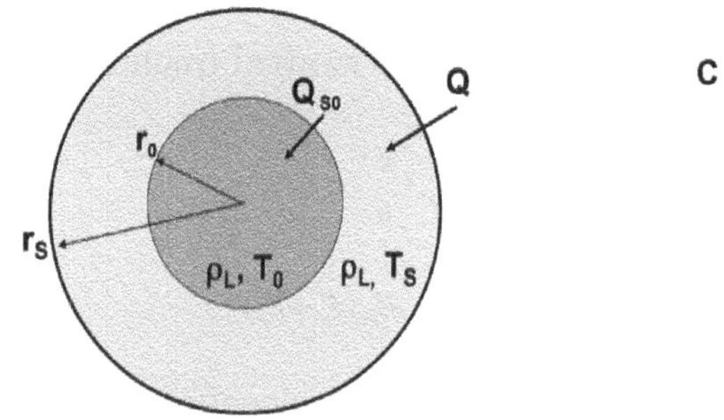

Figure 4.1. *Discrete two-layer model. The combustion gases serve for the evaporation at the level of the gas / liquid interface and pass on to the outer layer the quantity of heat Q. Similarly, that outer layer passes the quantity of heat Q_{S0} to the nucleus*

The ratio ε of the volume of the nucleus to that of the drop is supposed to be constant and is determined by a separate optimization calculation. The evaporation rate of the drop is always: $\dot{M} = \pi \rho \mathcal{D} Sh * D \ln(1 + B_M)$, in the case of relative velocity, where Sh is the Sherwood number, D the drop diameter and $\mathcal{D}$ the diffusion coefficient.

The system of equations therefore contains an extra equation [MAU 08]:

$$\frac{\partial \rho_L e_{t0}}{\partial t} + div(\rho_L e_{t0} \mathbf{u}) = n Q_{S0} \qquad [4.23]$$

with:

$$\rho_0 = \varepsilon \alpha \rho_L(T_0),\ e_{t0} = e(T_0) + |\mathbf{u}|^2/2\ ,\quad e(T_0) = e_{ref} + c_s(T_0)(T_0 - T_{ref})$$

The flux Q_{S0} represents the heat communicated from the outer layer of the drop to the central nucleus. It is determined by the linear relation:

$$Q_{S0} = M_0\, c_L\, \beta_0\, (T - T_0)$$

with M_0 being the mass of the central nucleus of the drop:

$$M_0 = \rho_L(T_0)\frac{\pi}{6} D_0^3 = \rho_L(T_0)\frac{\pi}{6} D^3\, \varepsilon$$

We have: $nQ_{S0} = \rho_L(T_0)\alpha\, c_L\, \beta_0\, (T - T_0)$.

NOTE.– The study of the exchange coefficient β_L brings into play an approximate relation between it and the ratio of volume ε: $\beta_L = 12\,\kappa_L \Big/ D^2\left(\varepsilon^{2/3} - \varepsilon\right)$, with $\kappa_L = k_L / \rho_L\, c_L$, in the case of quantities k_L, $\rho_L = \rho_0$, $c_L = c_s$, uniform and constant. This relation leads to: $Q_{S0} = 2\pi\, k_L\, D(T - T_0) \Big/ \left(\varepsilon^{-1/3} - 1\right)$.

The other equations are formally unchanged. The final system, therefore, for a single class of particles, is:

$$\begin{cases} \dfrac{\partial \rho}{\partial t} + div(\rho\,\mathbf{u}) = -n\,\dot{M} \\ \dfrac{\partial \rho\,\mathbf{u}}{\partial t} + div(\rho\,\mathbf{u} \otimes \mathbf{u}) = n\,\mathbf{F} - n\,\dot{M}\,\mathbf{u} \\ \dfrac{\partial \rho\, e_t}{\partial t} + div(\rho\, e_t\, \mathbf{u}) = n(\mathbf{F}\cdot\mathbf{u} + Q) - n\,\dot{M}\,(h_S + \dfrac{1}{2}|\mathbf{u}|^2) \\ \dfrac{\partial\, \rho_L e_{t0}}{\partial t} + div(\rho_L e_{t0}\mathbf{u}) = nQ_{S0} \\ \dfrac{\partial n}{\partial t} + div(n\,\mathbf{u}) = 0 \end{cases} \qquad [4.24]$$

4.1.6.2. *Case of the N-layer model*

The N-layer model is the model which needs to be implanted in CEDRE, of which the 2-layer case is only an example. Here, there are *N*-1 additional variables T_0 to T_{N-2} and *N*-1 ratios of volume ε_0 to ε_{N-2}, to be introduced in order to implement the *N*-layer model in the solver. The resulting equations are established by Mauriot and Prud'homme in [MAU 08]. This time, we write:

$$\rho = \sum_{i=0}^{N-2} \alpha_i \rho_L(T_i) + \alpha_{N-1} \rho_L(T) \qquad [4.25]$$

with:

$$\alpha = n\pi D^3/6,\ \varepsilon_0 = D_0^3/D^3$$

$$\varepsilon_i = \left(D_i^3 - D_{i-1}^3\right)\Big/D^3 \quad (\text{where } D_{N-1} = D)$$

$$\sum_{i=0}^{N-1} \varepsilon_i = 1, \alpha_i = \alpha\,\varepsilon_i$$

The specific density of the liquid is $\rho_L(T)$, a given function of the temperature, $T_{N-1} = T$ the temperature of the outer layer in contact with the gaseous mixture (called T_S in our evaporation models), T_i the temperature of the ith internal layer and T_0 the temperature of the inner nucleus. The ratios of the volume ε_0 to ε_{N-2} are supposed to be constant (but not necessarily equal) and are determined by a separate algorithm. The evaporation rate remains unchanged in relation to the 2-layer model: $\dot{M} = \pi\rho \mathcal{D} Sh^* D \ln(1+B_M)$. The system of equations includes *N*-1 additional equations from *i* = 0 to *i* = *N*-2:

$$\frac{\partial \rho_i e_{ti}}{\partial t} + \nabla\cdot(\rho_i e_{ti}\mathbf{u}) = nQ_{i-1,i} \qquad [4.26]$$

with:

$$\rho_i = \varepsilon_i \, \alpha \, \rho_L(T_i),\ e_{ti} = e(T_i) + |\mathbf{u}|^2/2\ ,\ e(T_i) = e_{ref} + c_s(T_i)(T_i - T_{ref})$$

$$Q_{i+1,i} = M_i \, c_L \, \beta_i \, (T_{i+1} - T_i)$$

with M_i being the mass of the ith internal layer, for $i = 0$ to N-1: $M_i = \rho_L(T_i)\frac{\pi}{6} D^3 \, \varepsilon_i$.

The flux $Q_{i+1,i}$ represents the heat communicated from the $(i + 1)^{th}$ layer to the ith layer ($i = 0$ to N-1). It is determined by the relation:

$$nQ_{i+1,i} = \rho_L(T_i)\,\alpha_i \, c_L \, \beta_i \, (T_{i+1} - T_i)$$

The other equations are formally unchanged. For a particular class of particles, the final system is:

$$\begin{cases} \dfrac{\partial \rho}{\partial t} + div(\rho \, \mathbf{u}) = -n\dot{M} \\ \dfrac{\partial \rho \, \mathbf{u}}{\partial t} + div(\rho \, \mathbf{u} \otimes \mathbf{u}) = n\,\mathbf{F} - n\dot{M}\,\mathbf{u} \\ \dfrac{\partial \rho \, e_t}{\partial t} + div(\rho \, e_t \, \mathbf{u}) = n(\mathbf{F}\cdot\mathbf{u} + Q) - n\dot{M}\,(h_S + \frac{1}{2}|\mathbf{u}|^2) \\ \dfrac{\partial \, \rho_i e_{ti}}{\partial \, t} + div(\rho_i e_{ti}\mathbf{u}) = nQ_{i+1,i} \\ \dfrac{\partial n}{\partial t} + div(n\,\mathbf{u}) = \vec{0} \end{cases} \quad [4.27]$$

4.2. Application to a spray flame

4.2.1. *Application of a minimum model to the study of the threshold of appearance of a pulsating flame*

An asymptotic study performed by Nicoli, Haldenwang and Suard [NIC 05, NIC 09a], summarized in [PRU 10], enabled us to analyze the stability of a spray flame in a poor mixture in the configuration shown in Figure 4.2, in the context of simplified hypotheses:

– the scales of combustion and of the spray are clearly separate such that $\ell_d << \ell_f$: the distance between drops is very small in relation to the length of the reaction–diffusion. We can show that this condition is equivalent to $Da << 1$, where the Damköhler parameter is equal to the ratio of the characteristic vaporization time to the chemical time;

– the medium is considered as a continuum with the three species: air, gaseous fuel and liquid fuel. We do not take account of the diffusion of the droplets;

– the latent heat is supposed to be null;

– in the gas, there is no heat diffusion, the external forces and viscosity are negligible, the static pressure can initially be taken to be constant, and we adopt Fourier's law for heat conductivity and Fick's law with a single coefficient for the diffusion. The Lewis number is equal to one, there is no chemical reaction and we are dealing with a mixture of perfect gases.

The dimensionless system to be resolved in these hypotheses is:

$$\begin{cases} \partial Y_l/\partial t + v \cdot \nabla Y_l = -\dot{W}_{vap} \\ \partial Y_g/\partial t + v \cdot \nabla Y_g = \Delta Y_g + \dot{W}_{vap} - \dot{W}_{chem} \\ \partial \theta/\partial t + v \cdot \nabla \theta = \Delta \theta + \dot{W}_{chem} \end{cases} \qquad [4.28]$$

with:

$$\theta(x,y,t) = (T - T_u)/(T_b - T_u), \quad Y_l = Y_l/(Y_{lu} + Y_{gu}), \quad t_{ref} = (D_{th})_b / U^2_{ZFK}$$

The subscripts u and b respectively represent unburned and burned gases, $x_{ref} = \frac{(D_{th})_b}{U_{ZFK}}$, $U_{ZFK} = \sqrt{\frac{2Le\,\lambda\,\rho_b^2\,Y_{O_2}\,B}{C_p\,\rho_f^2\,Ze^2}}e^{-E_a/2RT_b}$, $\dot{W}_{vap} = Y_l H(\theta-\theta_v)/Da$, $\dot{W}_{chem} = (Ze^2\,Y_g/2)e^{-Ze(1-\theta)}$, H is the Heaviside step function, Da is a Damköhler parameter equal to the ratio of the characteristic vaporization time to the chemical time: $Da = \tau_{vap}/\tau_{chim} = (\ell_D/\ell_f)^2$, ℓ_D the distance between the drops, and ℓ_f the diffusion–reaction length.

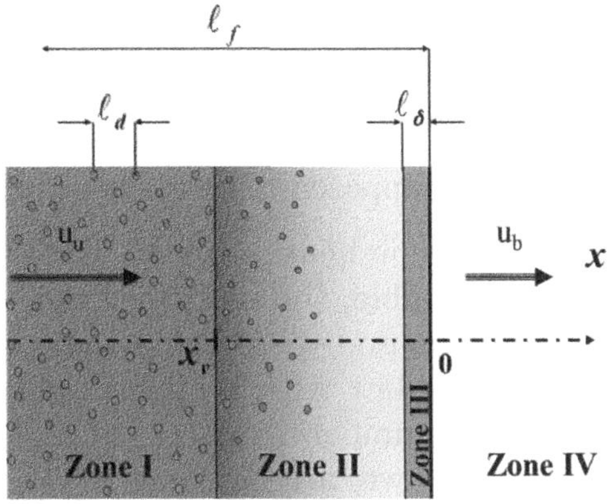

Figure 4.2. *Configuration of the 1D flame discussed by Nicoli et al. [NIC 05]. Zone I, with its border x_v of the start of vaporization. Zone II, with vaporization. Zone III, which is asymptotically thin, where the combustion reaction takes place. Zone IV where the burned gases are at equilibrium*

Asymptotic analysis is performed in all four zones in Figure 4.2. It gives the profiles of temperature and gas/liquid concentrations (Figure 4.3(a)). For each value of the Damköhler number, in the plane of the critical Zeldovich number[5] as a function of the wavenumber, the stability analysis gives us the curve of separation of the steady-state

5 The Zeldovich number: $Ze = E_a(T_b - T_u)/RT_b^2$ represents the reduced activation energy.

spray flame and the pulsating spray flame (see Figure 4.3(b)).

The authors observe that the presence of droplets decreases the threshold of the pulsating instabilities in comparison with the case of the gaseous premixed flame.

Theoretical and experimental studies demonstrate [ATZ 06][6] that the pulsating behavior is an intrinsic phenomenon of fog flames. It corresponds to an interaction between the evaporation front, which precedes the flame, and the chemical reaction zone. The configurations studied relate, in particular, to poor fogs for which the Damköhler number (the ratio of the vaporization time to the chemical time) is small. This pulsating regime is similar to a Hopf bifurcation (see Appendix), whose control parameter is the Zeldovich number (which can be interpreted as the ratio of the activation energy for the reaction to the kinetic energy of the burned gases).

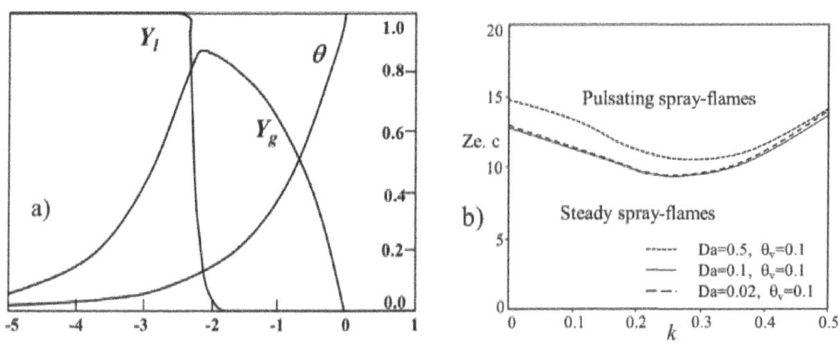

Figure 4.3. *a) 1D fog flame structure used by Nicoli et al. for $Da = 0.1$, $\theta_v = 0.1$. b) Neutral curve of the pulsating instability of a fog flame for different values of the Damköhler number, k is the wavenumber, Ze, c is the critical Zeldovich number [NIC 05]*

---

6 The authors note differences in velocity between the drops and the gas in the vicinity of the flame front. They deduce that these cause local variations in the richness of the mixture which, in turn, lead to oscillations in the velocity of the flame.

4.2.2. *Application to the study of the resonant action of an acoustic wave on a spray flame*

4.2.2.1. *Base equations*

The same authors applied their minimum model by subjecting the four-zone domain to a *temporal sinusoidal pressure variation* [NIC 09b, NIC 10], uniform over the whole of the domain of calculation of the flame:

$$p' = \varepsilon\, p_0 \sin \omega t$$

with: $p_0 = \rho_b T_b \left(c_p - c_v\right)$.

The equations are as follows:

$$\begin{cases} \partial Y_l / \partial t + v \cdot \nabla Y_l = -\dot{W}_{vap} \\ \partial Y_g / \partial t + v \cdot \nabla Y_g = (1/Le)\,\Delta Y_g + \dot{W}_{vap} - \dot{W}_{chem} \\ \partial \theta / \partial t + v \cdot \nabla \theta = \Delta \theta + \dot{W}_{chem} + \left(\alpha_p / \rho\right) \partial p / \partial t \end{cases} \qquad [4.29]$$

where $\alpha_p = \dfrac{c_p - c_v}{c_p} \dfrac{T_b}{T_b - T_u}$.

4.2.2.2. *Study of coupling*

First, the authors study the thermo-acoustic coupling for the parameters:

$$Da = 0.1,\ \theta_v = 0.1,\ Le = 1,\ \left(Y_l\right)_u = 1$$

For these values of the parameters, the natural oscillator appears beyond $Ze_c = 15.25$ with the natural pulsation.

To begin with, we are interested in the response of that fog flame to acoustic forcing "resonating at the threshold" (i.e. the acoustic pulsation is fixed at $\omega = \omega_c = 0.2827$) with an amplitude of the forcing $\varepsilon = 0.01$ for the values 15.25, 13 and 17 of the Zeldovich number (which increases with the percentage of nitrogen in the pre-mixture).

For $Ze_c = 15.25$, the acoustic wave greatly disturbs the fluctuation of the reaction rate (Figure 4.4), integrated over the domain (notated as Q'); indeed, this quantity is not sinusoidal (as it would be without the forcing), but rather resembles a relaxation signal, exactly like the velocity of the flame. However, it remains in phase with the pressure.

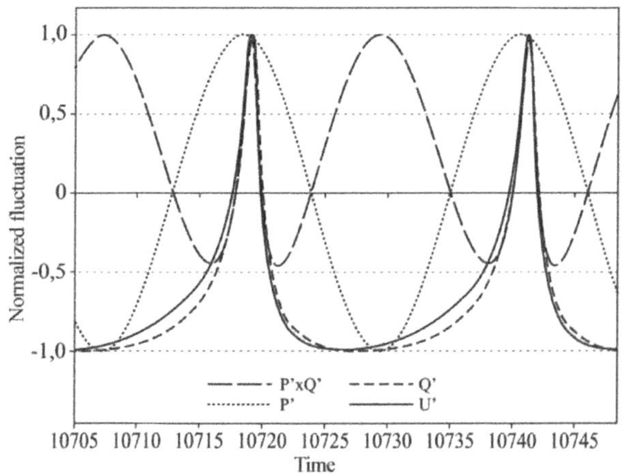

Figure 4.4. *Evolution over time of the acoustic pressure P', of the flame velocity U' and the energy release Q', and the acoustic growth rate* $P' \times Q'$ *of the fog flame subjected to an acoustic wave of force* $\varepsilon = 0.01$ *and pulsation* $\omega_c = \omega = 0.2827$ *with* $Ze = Ze_c = 15.25$ *(according to [NIC 09b], reproduced with kind permission from the authors)*

If we decrease *Ze* below the threshold *Ze*c, the natural pulsation will, in principle, be damped, but the forcing maintains an oscillation of slight amplitude. The phenomenon, however, remains unstable, because the average slope of the acoustic energy $\int_0^t P'Q'dt$ remains positive, although it is low.

If the Zeldovich number is taken, this time, above its critical value, the fluctuations in the velocity and the reaction are in phase opposition in relation to the pressure. The average slope of the acoustic energy becomes negative.

According to the Rayleigh criterion, this open-loop study shows that there would be no thermo-acoustic instability.

4.2.2.3. *Energy transfer to the acoustic wave*

Let us now consider the gains in energy received by the acoustic wave. Once again, we observe a resonant behavior: for each acoustic frequency, the gain is high when the natural oscillation frequency of the flame becomes similar. Indeed, when the acoustic frequency decreases, the maximum occurs for increasing Zeldovich numbers, which correspond to the expected threshold of the *Hopf instability* with precisely that frequency.

Thus, the transfer of combustion energy to the acoustic field is found to be maximal when the natural frequency of the pulsating flame and the acoustic frequency are in accordance.

Finally, the study of the gains obtained as a function of the amplitude of the forcing excitation shows that at a constant frequency of excitation, the gain is proportional to ε regardless of the Zeldovich number.

5

Radiative Phenomena

The objective of this chapter is, firstly, to lay down the definitions of the quantities playing a role in flows with radiative transfer, specify the hypotheses to be used for flows involved in aerothermochemistry, and provide the equations to use in the context of those hypotheses. The fundaments of the model are given in works such as the book by Modest [MOD 03] and an article by Tessé and Lamet [TES 11].

We shall then give a more-or-less complete discussion of a few applications, such as:

– hypersonic flow of atmospheric re-entry;

– the boundary layer with a diffusion flame over a flat plate with soot production resulting from an injection of fuel at the wall or the combustion of a solid plate (the classic Emmons problem will be recapped);

– the flow in an engine with combusting particles and radiative transfer.

Numerical methods will be mentioned with regard to the applications. We discuss a frequent usage of the Monte Carlo method [TES 01, TES 02, TES 04].

We note significant coupling between the flow and the radiative transfer in the aforementioned cases. Also, coupling is frequently observed between radiative transfer and the exchanges at the walls. For example, in solid aluminum propellant engines, there is an effect on the ignition because of the heat transmitted by radiative transfer to the surface.

In re-entering bodies, the shockwave causes the ionization of the air at high temperature. The radiative emission increases the heat flow toward the surface; that heat flow becomes essentially radiative for high-velocity re-entries. The resulting imbalance of the internal modes of energy of the plasma causes their relaxation and may impact on this flux. However, the radiative processes often have a significant imbalance in terms of the Boltzmann's distribution (the populating of the radiative levels).

Note that our knowledge and models of the radiative transfer have benefited from the expertise acquired in other scientific domains, such as astrophysics – particularly in relation to stars' atmospheres – but intense research is still being done in this area throughout the world.

5.1. Basic values and fundamental relations in radiative transfer

The important values to determine in order to introduce the effects of *radiative heat transfer* in flows are the flux vector of radiative transfer and the radiative power.

5.1.1. *Definitions*

We define the *monochromatic intensity* $I_\nu(M,\mathbf{u})$ (intensity of monochromatic radiative transfer of wavenumber ν) at any given point M in the volume or on its surface in direction $\mathbf{u}$. Depending on the case, it may be an emitted radiative transfer or a received incident radiative transfer.

The corresponding intensities are $I_\nu^{emi}(M,\mathbf{u})$ and $I_\nu^{inci}(M,\mathbf{u})$. *At complete equilibrium* of a radiating cavity, these two values are equal to a function of the temperature alone (black-body equilibrium) called a Planck function $I_\nu^0(T)$:

$$I_\nu^0(T)=\frac{2hc_0^{\,2}\nu^3}{e^{hc_0\nu/kT}-1} \qquad [5.1]$$

where h is Planck's constant, k Boltzmann's constant and c_0 the speed of light in a vacuum.

At local thermodynamic equilibrium (LTE)[1], the monochromatic intensity emitted at a point B is written as:

$$I_\nu^{emi}(B,\mathbf{u})=\varepsilon_\nu(B,\mathbf{u})L_\nu^0(T_B) \qquad [5.2]$$

where $\varepsilon_\nu(B,\mathbf{u})$ is monochromatic emissivity in the direction $\mathbf{u}$.

We shall see later on how to determine the intensity along a light beam by solving the radiative transfer equation. Before we do this, however, it is important to define two other values: the radiative transfer flux vector and the radiated power.

The *radiative transfer flux vector* is, at every point M of the volume or the surface:

$$\mathbf{q}^R(M)=\int_0^\infty\int_{4\pi} I_\nu(M,\mathbf{u})\,\mathbf{u}\,d\Omega\,d\nu \qquad [5.3]$$

$d\Omega$ being the solid angle element.

1 Hereinafter, we shall accept the hypothesis of LTE, as do the "CFD" codes used by ONERA [TES 11], but in reality this is not the case with atmospheric re-entry.

At a point B, the *local radiative flux from a wall* is simply $-\mathbf{q}^R(B)\cdot\mathbf{n}_B$, where $\mathbf{n}_B$ is the unitary normal to the surface at B orientated from the surface to the volume or, more specifically:

$$\phi^R(B) = -\int_0^\infty \int_{4\pi} I_\nu(B,\mathbf{u})\,\mathbf{u}\cdot\mathbf{n}_B\, d\Omega\, d\nu \qquad [5.4]$$

The radiative power at a point M is equal to $-(\nabla\cdot\mathbf{q}^R)(M)$, so:

$$P^R(M) = -\int_0^\infty \int_{4\pi} \frac{\partial I_\nu(s,\mathbf{u})}{\partial s}\, d\Omega\, d\nu \qquad [5.5]$$

where s is the abscissa value along the light beam passing through M.

These quantities are global for all of the frequencies and solid angles.

The radiative flux $\phi^R(B)$ can be deconstructed into two parts: the incident radiative transfer originating within the volume and hitting the wall, and the radiative transfer emitted by the wall itself:

$$\phi^R(B) = \int_0^\infty \int_{2\pi(\mathbf{u}\cdot\mathbf{n}_B<0)} I_\nu^{inci}(B,\mathbf{u})\,|\mathbf{u}\cdot\mathbf{n}_B|\, d\Omega\, d\nu - \int_0^\infty \int_{4\pi} I_\nu^{emi}(B,\mathbf{u})\,|\mathbf{u}\cdot\mathbf{n}_B|\, d\Omega\, d\nu \qquad [5.6]$$

5.1.2. *Radiative Transfer Equation (RTE)*

In order to determine the radiative flux and power, it is necessary to know the intensity of the directional monochromatic radiative transfer $I_\nu(M,\mathbf{u})$ along a light beam, at any given abscissa value s. Therefore, we need to establish the radiative energy balance. Here, we give the result obtained in the steady-state, isotropic case[2]:

$$\begin{aligned}\frac{\partial I_\nu(s,\mathbf{u})}{\partial s} &= \eta_\nu(s) + \tfrac{\sigma_\nu(s)}{4\pi}\int_0^\infty\int_{4\pi} I_\nu(s,\mathbf{u}')\Phi_\nu(s,\mathbf{u}'\rightarrow\mathbf{u})d\Omega' \\ &-\left[\kappa_\nu(s)+\sigma_\nu(s)\right]I_\nu(s,\mathbf{u})\end{aligned} \quad [5.7]$$

In this equation, $\eta_\nu(s)$, $\kappa_\nu(s)$ and $\sigma_\nu(s)$ are, respectively the monochromatic coefficients of emission, absorption and scattering at s; $\Phi_\nu(s,\mathbf{u}'\rightarrow\mathbf{u})$ is the scattering phase function, representing the angular distribution of the intensity diffused, which, for a beam propagating in the direction $\mathbf{u}'$, leads us to the probability of its being scattered in the direction $\mathbf{u}$ at abscissa s: $\Phi_\nu(s,\mathbf{u}'\rightarrow\mathbf{u})d\Omega'/4\pi$. The first terms on the right-hand side of equation [5.7] correspond to the gain by the emission and the incoming scattering; the third corresponds to the losses by absorption and scattering of light.

By integrating over an optical thickness between $s'=s_{B_0}$ (on the surface) and $s'=s$ (within the volume), we obtain the *integral form of the radiative transfer equation* (or RTE):

$$I_\nu(s,\mathbf{u}) = I_\nu^{leav}(s_{B_0},\mathbf{u})\tau_\nu(s_{B_0},s) + \int_{s_{B_0}}^{s} S_\nu(s',\mathbf{u})\tau_\nu(s',s)ds' \quad [5.8]$$

2 On this point, Tessé and Lamet [TES 11] note that this case covers a large majority of the flows studied at ONERA, and justify the hypotheses chosen by the argument that the characteristic time of propagation of the radiation is very slight in relation to the characteristic times of the other physical phenomena.

with the source function:

$$S_\nu(s',\mathbf{u}) = \eta_\nu(s') + \frac{\sigma_\nu(s')}{4\pi}\int_{4\pi} I_\nu(s',\mathbf{u}')\Phi_\nu(s',\mathbf{u}'\rightarrow\mathbf{u})\,d\Omega'$$

and the monochromatic transmissivity:

$$\tau_\nu(s',s) = \exp\left[-\int_{s'}^{s}\left(\kappa_\nu(s'')+\sigma_\nu(s'')\right)ds''\right]$$

The monochromatic intensity $I_\nu^{leav}\left(s_{B_0},\mathbf{u}\right)$, leaving the wall in direction **u**, is the sum of the emitted and reflected intensities, so, in the case of a mirror:

$$I_\nu^{leav}(B_0,\mathbf{u}) = \varepsilon_\nu(B_0,\mathbf{u})I_\nu^0\left(T_{B_0}\right) + \left[1-\varepsilon_\nu(B_0,\mathbf{u})\right]I_\nu^{inci}(B_0,\mathbf{u}^*) \qquad [5.9a]$$

and, in the case of a diffusively reflective surface:

$$I_\nu^{leav}(B_0,\mathbf{u}) = \varepsilon_\nu(B_0,\mathbf{u})I_\nu^0\left(T_{B_0}\right) + \left(\frac{1-\varepsilon_\nu(B_0,\mathbf{u})}{\pi}\int_{2\pi(\mathbf{u}\cdot\mathbf{n}_{B_0}<0)} I_\nu^{inci}(B_0,\mathbf{u})\left|\mathbf{u}'\cdot\mathbf{n}_{B_0}\right|d\Omega'\right) \qquad [5.9b]$$

5.1.3. *Radiative flux and power*

From the above results, we can deduce the final forms of the radiative flux and power from the surface:

$$\begin{aligned}\phi^R(B) = &\int_0^\infty\int_{2\pi(\mathbf{u}\cdot\mathbf{n}_B<0)} \alpha_\nu(B,\mathbf{u})I_\nu^{leav}(B_0,\mathbf{u})\tau_\nu\left(s_{B_0},s_B\right)\left|\mathbf{u}\cdot\mathbf{n}_B\right|d\Omega\,d\nu + \\ &\int_0^\infty\int_{2\pi(\mathbf{u}\cdot\mathbf{n}_B<0)} \alpha_\nu(B,\mathbf{u})\int_{s_{B_0}}^{s_B} S_\nu(s',\mathbf{u})\tau_\nu(s',s_B)\left|\mathbf{u}\cdot\mathbf{n}_B\right|d\Omega\,d\nu - \\ &\int_0^\infty I_\nu^0(T_B)\int_{2\pi(\mathbf{u}\cdot\mathbf{n}_B<0)} \varepsilon_\nu(B,\mathbf{u})\left|\mathbf{u}\cdot\mathbf{n}_B\right|d\Omega\,d\nu\end{aligned} \qquad [5.10]$$

where the first terms on the right-hand side of the equation correspond to absorption and the final term is due to light emission, and:

$$P^R(M) = -\int_0^\infty \kappa_\nu(M) \int_{4\pi} I_\nu(M,\mathbf{u})\, d\Omega\, d\nu - 4\pi \int_0^\infty \eta_\nu(M)\, d\nu \qquad [5.11]$$

In equation [5.11], the first term on the right-hand side corresponds to absorption and the second to emission.

5.1.3.1. *Monochromatic absorption coefficient*

For a suspension, the *absorption coefficient* can be expressed as the sum of the contributions due to the gas and the particles:

$$\kappa_\nu(M) = \kappa_\nu^{gas}(M) + \kappa_\nu^{part}(M) \qquad [5.12]$$

The contribution of the gas molecules often has a negligible effect on the radiative transfer in comparison to the contribution of the particles, but the radiative state of the particles depends on their interactions with the gas.

The transmissivity is written as follows, using equation [5.12]:

$$\begin{aligned}\tau_\nu(s',s) &= \exp\left[-\int_{s'}^{s} \kappa_\nu^{gaz}(s'')\, ds''\right] \exp\left[-\int_{s'}^{s} \left(\kappa_\nu^{part}(s'') + \sigma_\nu(s'')\right) ds''\right] = \\ &= \tau_\nu^{gaz}(s',s)\, \tau_\nu^{part}(s',s)\end{aligned} \qquad [5.13]$$

Additionally, if we cannot use equation [5.11] directly, we use equation [5.8], which gives us:

$$\begin{aligned}&P^R(M) = \\ &\int_0^\infty \kappa_\nu(M) \int_{4\pi} \Bigg[I_\nu(M,\mathbf{u})\, \tau_\nu\left(s_{B_0}, s_M\right) + \int_{s_{B_0}}^{s_M} S_\nu(s',\mathbf{u})\, \tau_\nu(s',s_M)\, ds' \Bigg] d\Omega\, d\nu \quad \text{absorption} \\ &-4\pi \int_0^\infty \eta_\nu(M)\, d\nu \quad \text{emission}\end{aligned} \qquad [5.14]$$

By multiplying equation [5.12] by equation [5.13], we deduce:

$$\kappa_\nu(s)\,\tau_\nu(s',s) = \kappa_\nu^{gas}(s)\,\tau_\nu^{gas}(s',s)\,\tau_\nu^{part}(s',s) + \kappa_\nu^{part}(s)\,\tau_\nu(s',s) = -\frac{\partial \tau_\nu^{gas}(s',s)}{\partial s}\tau_\nu^{part}(s',s) + \kappa_\nu^{part}(s)\,\tau_\nu(s',s) \quad [5.15]$$

Using equations [5.15] and [5.14], we obtain:

$$P^R(M) = \underbrace{\int_0^\infty \int_{4\pi} \left[-\frac{\partial \tau_\nu^{gaz}(s_{B_0},s)}{\partial s}\bigg|_{s=s_M} \tau_\nu^{part}(s_{B_0},s_M) + \kappa_\nu^{part}(s_M)\,\tau_\nu\left(s_{B_0},s_M\right)\right] I_\nu^{leav}(B_0,\mathbf{u})\,d\Omega\,d\nu}_{\text{absorption}} + \underbrace{\int_0^\infty \int_{4\pi} \int_{s_{B_0}}^{s_M} \left[-\frac{\partial \tau_\nu^{gaz}(s',s)}{\partial s}\bigg|_{s=s_M} \tau_\nu^{part}(s',s_M) + \kappa_\nu^{part}(s_M)\,\tau_\nu(s',s_M)\right] S_\nu(s',\mathbf{u})\,ds'\,d\Omega\,d\nu}_{\text{absorption}} - \underbrace{4\pi\int_0^\infty \eta_\nu(M)\,d\nu}_{\text{emission}} \quad [5.16]$$

Equation [5.16] is useful if we formulate the properties of the gas model in terms of transmissivity, because the *gas absorption coefficient does not come into play.*

5.1.3.2. *Monochromatic emission coefficient*

If we accept the hypothesis of LTE (case of combustion, but rarely the case for re-entry flows), we obtain the following for the monochromatic emission coefficient involved in equation [5.7]:

$$\eta_\nu(M) = \kappa_\nu(M)\, I_\nu^0\left(T_M^{gaz}\right) + \sum_{k=1}^{N_{part}} \kappa_\nu^{p,k}(M)\, I_\nu^0\left(T_M^{p,k}\right) \quad [5.17]$$

N_{part} being the number of particle sizes considered in the simulation, which may be in a state of disequilibrium with the gas.

If we discount the influence of the particles, the radiative power given by equation [5.16] is simplified, becoming:

$$P^R(M) = -\int_0^\infty \int_{4\pi} \underbrace{\left.\frac{\partial \tau_\nu^{gas}\left(s_{B_0}, s\right)}{\partial s}\right|_{s=s_M}}_{\text{absorption}} \tau_\nu^{part}\left(s_{B_0}, s_M\right) I_\nu^{leav}\left(B_0, \mathbf{u}\right) d\Omega\, d\nu - \quad [5.18]$$

$$\int_0^\infty \int_{4\pi} \int_{s_{B_0}}^{s_M} \underbrace{\left.\frac{\eta_\nu(s')}{\kappa_\nu(s')} \frac{\partial^2 \tau_\nu^{gas}(s',s)}{\partial s\, \partial s'}\right|_{s=s_M}}_{\text{absorption}} ds'\, d\Omega\, d\nu - \underbrace{4\pi \int_0^\infty \eta_\nu(M)\, d\nu}_{\text{emission}}$$

where the ratio η_ν / κ_ν can be replaced by function [5.1].

5.1.4. *Involvement of radiative heat transfer in the equations of aerothermochemistry*

The fundamental equations of reactive flows discussed in Volumes 1 and 2 of this series [PRU 12, PRU 13] are modified when it comes to the total energy balance equation, $e_t = e + v^2/2$, by the intervention of a source term equal to the radiated power. This term plays a part in the homogeneous phase, as well as at the interface and at the walls, but in a different way. In a multi-phase medium, we need to take account of the amounts of power emitted and received by each of the phases present. At the high temperatures which manifest in radiative flows, numerous couplings take place, with various phenomena in a state of disequilibrium, such as the internal energy modes (thermal disequilibrium), the chemical reactions generally interacting with the modes of vibration of the molecules (on this subject, see [PRU 12, PRU 13] and also Aurélien Guy's doctoral thesis [GUY 13]).

It is also necessary to take into consideration the effects of turbulence.

5.1.4.1. *Case of a one-dimensional homogeneous gaseous flow*

The simplified one-dimensional steady-state system of equations used by Allouche *et al.* [ALL 13] – to study the fluid flow, downstream the frontal shockwave, during high-velocity re-entry of bodies into the atmosphere – is as follows:

$$\begin{cases} d(\rho u)/dx=0, \quad d\left(p+\rho u^2\right)/dx=0, \quad d(\rho_\alpha u)/dx=\dot{W}_\alpha, \\ \dfrac{d}{dx}\left[u\left(\rho(e+k)+p\right)+q^{rad}\right]=0, \quad \dfrac{d}{dx}\left(\rho_\alpha e_{v,\alpha} u\right)=\dot{W}_{v,\alpha}+\dot{W}_\alpha e_{v,\alpha}, \\ dI_v/dx=\varepsilon_v-I_v\,\kappa_v \end{cases} \quad [5.19]$$

In this differential system, $(\rho, u, p \text{ and } \rho_\alpha)$ are the classic variables respectively denoting the density, velocity, pressure and partial density of the species α. The total mass energy is the sum of the internal energy and the kinetic energy. The radiative flux is denoted by q^{rad}, the vibration energy of the diatomic species α is $e_{v,\alpha}$ and the monochromatic intensity for the wavenumber v is I_v. The production terms are: $\dot{W}_\alpha$ for the chemistry, and $\dot{W}_{v,\alpha}$ for the vibration energy. The values η_v and κ_v are the monochromatic emission and absorption coefficients. As we envisage a mixture of perfect gases with a temperature of the electrons different to that which is common to the ions and to the neutral particles, we write the pressure:

$$p=\sum_{\alpha\neq e}\rho_\alpha RT/\mathcal{M}_\alpha+\rho_e RT/\mathcal{M}_e$$

We shall look again at this problem in section 5.2. Let us simply note here that the dissipative phenomena of diffusion of the species, heat conduction and viscosity are not taken into account, and that we are dealing with a homogeneous gaseous mixture. The heterogeneity of that flow lies in the presence of the wall of the re-entering body in the vicinity of the flow stop point.

5.1.4.2. *Case of a three-dimensional homogeneous gaseous flow*

Lamet [LAM 09] writes the set of 3D balance equations in a more general scenario:

$$\left.\begin{array}{ll}
\partial \rho_{\alpha;i}/\partial t+\nabla\cdot\left(\rho_{\alpha;i}\mathbf{v}\right)=\nabla\cdot\left(\rho D_{\alpha;i}\nabla Y_{\alpha;i}\right)+\dot{W}_{\alpha;i} & (a)\\
\partial(\rho\mathbf{v})/\partial t+\nabla\cdot(p\mathbf{1}+\rho\mathbf{v}\otimes\mathbf{v})=\nabla\cdot(2\mu\mathbf{D}) & (b)\\
\partial\left[\rho(e_e+k_e)\right]/\partial t+\nabla\cdot\left\{\mathbf{v}\left[\rho_e(e_e+k_e)+p_e\right]\right\}-\mathbf{v}\cdot\nabla p_e=\nabla\cdot(\lambda_e\nabla T_e)+ & \\
\nabla\cdot(\rho_e h_e D_e\nabla Y_e)+\dot{W}_e^{e-heavy}+\sum_r\dot{W}_e^{coll,r}+\sum_p\dot{W}_e^{rad,p} & (c)\\
\partial\left[\rho(e+k)\right]/\partial t+\nabla\cdot\left[\rho\mathbf{v}(h+k)\right]=\nabla\cdot(\lambda\nabla T)+\nabla\cdot(\lambda_e\nabla T_e)+ & \\
\nabla\cdot\left(\rho\sum_{\alpha,i}D_{\alpha,i}h_{\alpha,i}\nabla Y_{\alpha,i}\right)+\nabla\cdot(2\mu\mathbf{v}\cdot\mathbf{D}) & (d)\\
dI_\nu(s,\mathbf{u})/ds=\eta_\nu(s)-\kappa_\nu(s)I_\nu(s,\mathbf{u}) & (e)
\end{array}\right\} \quad [5.20]$$

We recognize the balance equations for (a) the mass of the species α, (b) the momentum of the mixture, (c) the total energy of the free electrons, (d) the total energy of the mixture, and (e) the radiative energy.

The main simplifying hypotheses made here are local electrical neutrality, negligible viscosity of volume, Fourier laws for the conduction and Fick laws for the diffusion in the gaseous mixture, and with the free electrons. We also assume a classic approximation to evaluate the energy involved in the electrical field, with the term $\mathbf{v}\cdot\nabla p_e$, and a mixture of perfect gases at two different temperatures.

J.-M. Lamet quite rightly points out that the system of equations [5.20] is never resolved exactly as it is in the literature, in view of the very high number of species (α, i) needing to be taken into account, not to mention the lack of data (reaction constants, diffusion coefficients, etc.) to simulate the transport phenomena and the reactional processes taking place between particles. To this, we must also add interaction with the radiative transfer, which is difficult to deal with, both because of its non-localized nature and because of the large number of processes which it involves [LAM 09].

5.1.4.3. *Case of a flow with soot formation and radiative transfer*

The problem examined by Dorey [DOR 12] for his doctoral thesis from ONERA is that of a flow with coupled phenomena between combustion, soot formation and radiative transfers, in the combustion chambers of gas turbines. The system of equations adopted for the gaseous mixture is:

$$\begin{cases} \rho\, dY_\alpha/dt + \nabla\cdot\mathcal{J}_\alpha = \dot{W}_\alpha + \dot{W}_\alpha^d \\ \rho\, d\mathbf{v}/dt + \nabla\cdot(\mathbf{\Pi} + p\mathbf{1}) = \dot{W}_{mt} \\ \rho\, de_t/dt + \nabla\cdot\mathcal{J}_{E_t} = -\nabla\cdot\left[(\mathbf{\Pi} + p\mathbf{1})\otimes\mathbf{v}\right] + \dot{W}_{E_t} \end{cases} \qquad [5.21]$$

The notations used are those typically employed for homogeneous fluids:

$$\rho, Y_\alpha, \mathcal{J}_\alpha, \dot{W}_\alpha, \mathbf{v}, \mathbf{\Pi}, p, e_t, \mathcal{J}_{E_t}$$

respectively for the average density, the mass fraction of the species α, the diffusion flux of that species (we can also write this as $\mathbf{J}_{D\alpha}$ instead of $\mathcal{J}_\alpha$), the barycentric velocity, the viscous pressure tensor, thermodynamic pressure, total mass energy and total energy transfer flux $(e+k)$, which is the sum of the internal energy and the kinetic energy. The production $\dot{W}_\alpha^d$ is due to the dispersed phase; $\dot{W}_{mt} \cong \dot{W}_{mt}^d$ corresponds to the volume forces – primarily those of the dispersed phase – with the influence of gravity being discounted (see Table 4.2 for the non-radiative case); finally, $\dot{W}_{E_t} = \dot{W}_{rad} + \dot{W}_{E_t}^d$, with the first term stemming from the source of radiative transfer, and the second from the dispersed phase. Readers can refer to Table 4.2 and section 4.1.4 to express $\dot{W}_\alpha^d$ in the case of vapor and in that of other gaseous species, as well as for $\dot{W}_{mt}$ and $\dot{W}_{E_t}^d$.

5.1.4.4. *Interface with radiative transfer*

To solve the problems of flow with radiative transfer, we need to be able to write the boundary conditions. In the case of the re-entering body discussed in section 5.1.4.1, we look at the surface of the re-entering body in the vicinity of the stagnation point of the flow. In the case of a combustion chamber, we look at the walls. The walls may be ablative in a solid-propellant thruster or on the surface of a re-entering body. At the scale of a particle, such as soot particles or material undergoing pyrolysis, the surface is a radiating interface.

Therefore, terms are added to the classic balance equations at the interface without capillarity in the absence of radiative transfer. The total energy balance is written here as:

$$\left[q_\perp + q_\perp^{rad} + (\mathbf{v} - \mathbf{V}_S) \cdot \mathbf{P} \cdot \mathbf{N} + \rho e_t u_\perp\right]_-^+ = \dot{W}_{E_t a} \qquad [5.22]$$

where $u_\perp = v_\perp - V_{S\perp}$ is the normal relative velocity in relation to the interface, $q_\perp^{rad} = -\mathbf{q}^{rad} \cdot \mathbf{N}$ is the radiative flux defined by equation [5.4] and refined with equation [5.6], with $\mathbf{N}$ being orientated toward the gas, and $\dot{W}_{E_t a}$ is a source of total energy per unit area and time at the interface, stemming from the contribution of the condensed phase and possibly that of radiative transfer. Finally: $\dot{W}_{E_t a} = \dot{W}^d_{E_t a} + \dot{W}_{rad\,a}$.

5.1.5. *Turbulence–radiation interaction (TRI)*

Here, we look at the contribution made by Tessé and Lamet [TES 11]. Other references are cited throughout the text for readers wishing to go into greater depth about this tricky issue, where turbulence interacts not only with the particles of the suspension and the chemical reactions, but also with the radiative heat transfer.

With the RANS methods [HIN 75] or LES (for subgrids, because they only involve temporal averages), it is necessary to know the relations governing the correlations.

[TES 11] begin by seeking to establish simplified models. For example, for a sooty flame, we can make the hypothesis of LTE with the absence of diffusion of radiative transfer.

In the absence of TRI, the RTE becomes:

$$\frac{\partial I_\nu(s,\mathbf{u})}{\partial s} = \kappa_\nu\left(\langle T(s)\rangle, \langle p(s)\rangle, \langle f_\nu(s)\rangle, \langle X_i(s)\rangle\right)\left[I^0_\nu\left(\langle T(s)\rangle\right) - I_\nu(s,\mathbf{u})\right] \qquad [5.23]$$

where $\langle T(s)\rangle, \langle p(s)\rangle, \langle f_\nu(s)\rangle, \langle X_i(s)\rangle$ are, respectively, the temporal averages of the temperature, pressure, volume fraction of soot and molar fractions of the species contributing to the radiative transfer.

If we do take account of TRI, and also use temporal averages for the radiative properties, the RTE to be considered becomes:

$$\frac{\partial \langle I_\nu(s,\mathbf{u})\rangle}{\partial s} = \left\langle \kappa_\nu\, I^0_\nu(s)\right\rangle - \left\langle \kappa_\nu I_\nu(s,\mathbf{u})\right\rangle \qquad [5.24]$$

Equation [5.24] shows that two correlations need to be considered:

– between the fluctuations in the absorption coefficient and those of the equilbrium intensity, which depend only on the local values T, p, f_v, X_i;

– between the fluctuations in the absorption coefficient and those of the radiative intensity, which is more complicated because it involves all the fluctuations in the meshes traversed by the optical path. For this reason, TRI is only partially taken into account, accepting that:

$$\langle \kappa_\nu(s) I_\nu(s,\mathbf{u}) \rangle \approx \langle \kappa_\nu(s) \rangle \langle I_\nu(s,\mathbf{u}) \rangle \qquad [5.25]$$

This is what we call the optically thin fluctuation approximation (OTFA), which is valid if and only if: $\langle \kappa_\nu(s) \rangle \ell << 1$, where ℓ is the dimension characteristic of the large turbulent eddies. It proves difficult to evaluate the error which results from this approximation.

Many authors do not employ the OFTA hypothesis, and consider the complete equation [5.20] (on this subject, see the study by Coelho [COE 07]). TRI is still very much a current research topic. Recent studies [DES 08, ROG 10], cited by Tessé and Lamet, use DNS (*Direct Numerical Simulation*) to model the effects of the fluctuations at subgrid scale on radiative transfer in LES (*Large Eddy Simulation* – see [SAG 06, POI 05, PRU 13]).

5.1.6. *Modeling of the radiative properties of gases*

For the radiative properties of gases, we use band models: i.e. we discretize the spectrum into intervals – the bands – and take the averages of the radiative properties in those bands. It is ensured that the Planck function $I_\nu^0(T)$ is constant throughout each band in the temperature interval in question.

The models are discussed in detail in Lamet's thesis [LAM 09], so here, we shall simply give a summary presentation of those mentioned in [TES 11] and used in the numerical codes at ONERA, which deal primarily with three applications which we shall look at later: atmospheric re-entry, role of soot and combustion of solid propellants.

The role of the spectral correlations[3] is important in the choice of the models to use. When the spectral correlations play a significant part, we need to consider *CK* (Correlated-K) *models* [TES 11]. We shall suppose, for the purposes of our discussion here, that this is not the scenario with which we are presented.

Rather, we choose to use *statistical narrow-band (SNB) models*, which are discussed in Lamet's thesis.

5.1.6.1. *Uniform medium*

5.1.6.1.1. Absorption coefficient and overlap parameter

In the case of a *uniform medium*, the transmissivity of a gas column of length l, which is homogeneous and isothermal, is written as:

$$\overline{\tau}^{\Delta\nu}(l) = \int_{\Delta\nu} e^{-xplk_\nu} d\nu \Big/ \Delta\nu = e^{-\overline{W}/\delta} \quad [5.26]$$

3 Spectral correlations occur when the emission takes place at the same position within the spectral as the absorption. If this is truly the case, then regardless of the value of the wavenumber ν within the range $(\nu, \nu+\Delta\nu)$, the resulting monochromatic intensity is null. The reality is often more complex. However, there are certain circumstances under which these correlations are negligible.

where $\delta = \Delta\nu/N$ is the average spacing of the N line positions of thickness $\Delta\nu$ and $\overline{W} = (1/N)\sum_{r=1}^{N}\int_{-\infty}^{+\infty}\left(1-e^{-\kappa_\nu' l}\right)d\nu$ is the equivalent black line width, with κ_ν' being the contribution of the rth line to the absorption coefficient, x the mass fraction of the species and p the pressure of the column.

The analytical expressions of $\overline{W}/\delta$ depend solely on two parameters: $\overline{k}$ and $\overline{\beta}$.

The parameter $\overline{k}$ is a reduced average absorption coefficient, linked to the average intensity of line $\overline{S} = \int S\,P(S)\,dS$ by the relation: $\overline{k} = \overline{S}/\delta$.

The parameter $\overline{\beta}$ characterizes the degree of overlap, and is proportional to the ratio $\gamma_0^{L/D}/\delta$, where L/D means "Lorentz or Doppler" (see section 5.1.6.1.2).

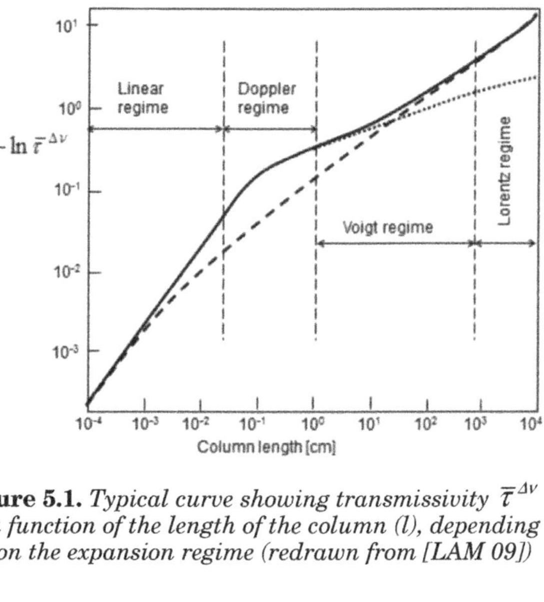

Figure 5.1. *Typical curve showing transmissivity $\overline{\tau}^{\Delta\nu}$ as a function of the length of the column (l), depending on the expansion regime (redrawn from [LAM 09])*

Generally, we can use the expression given by Ludwig [LUD 73]:

$$\overline{W_\nu}/\delta = x\,p\,l\,\bar{k}\sqrt{1-\Omega^{-1/2}} \qquad [5.27]$$

where:

$$\Omega = \left[1-\left(\frac{1}{x\,p\,l\,\bar{k}}\frac{\overline{W_D}}{\delta}\right)^2\right]^{-2} + \left[1-\left(\frac{1}{x\,p\,l\,\bar{k}}\frac{\overline{W_L}}{\delta}\right)^2\right]^{-2} - 1$$

A typical example is given by the growth curve in Figure 5.1, which represents the value $\left(-\ln\bar{\tau}^{\Delta\nu}\right)$ as a function of the length l of the column.

5.1.6.1.2. Spectral profiles

The spectral profile of a *Lorentz line* is:

$$f(\nu-\nu_c) = (\gamma_L/\pi)\Big/\left[\gamma_L^{\,2}+(\nu-\nu_c)^2\right]$$

which gives us: $\bar{k} = \bar{S}/\delta$, $\bar{\beta}_L = \bar{\gamma}_L/\bar{\delta}$, where $\bar{S}$ and $\bar{\gamma}_L$ are the intensity and the average width of the bands and $\bar{\delta}$ is the spacing parameter:

$$\bar{\delta} = \delta\,\bar{S}\,\bar{\gamma}_L\Big/\left[(1/N)\sum_{i=1}^{N}\gamma_{Li}S_i\right]^2$$

N being the number of lines. The parameter $\bar{k}$ is a sort of reduced mean absorption coefficient; the parameter $\bar{\beta}_L$ is an indicator of the overlap of the bands – it is larger when the overlap is greater. In the extreme case, we can write:

$$\text{if } x\bar{k}\,pl \ll \bar{\beta}_L \;\Rightarrow\; -\ln\bar{\tau}^{\Delta\nu} = x\bar{k}\,pl;\ \text{if } x\bar{k}\,pl \gg \bar{\beta}_L$$
$$\Rightarrow\; -\ln\bar{\tau}^{\Delta\nu} = 2\sqrt{x\bar{k}\,pl\,\bar{\beta}_L}$$

For a *Doppler line*:

$$f(\nu-\nu_c)=\left(\sqrt{\ln 2}/\gamma_D\sqrt{\pi}\right)\exp\left[-(\nu-\nu_c)^2\ln 2/\gamma_D{}^2\right]$$

Finally, a *Voigt line* has no analytical formulation and results from a combination between *Lorentz* and *Doppler*.

5.1.6.2. *Non-uniform medium*

In the case of a *non-uniform medium*, we use the Curtis-Godson approximation, considering that the behavior of the growth curve is the same as that of a uniform column, on condition that the abscissa value considered is the optical path $u=\int_0^l x(s)\,p(s)\,ds$.

The average transmissivity is therefore:

$$\overline{\tau}^{\Delta\nu}(l)=e^{-\overline{W}_\nu^*/\delta} \qquad [5.28]$$

The value $\overline{W}_\nu^*/\delta$ is given by equation [5.28] where we replace $x\,p\,k\,l$ with $\overline{k^*u}=\int_0^l x(s)\,p(s)\,\overline{k}(s)\,ds$, and $\overline{W_L}/\delta$ and $\overline{W_D}/\delta$ with $\overline{W_L^*}/\delta$ and $\overline{W_D^*}/\delta$.

With this approximation, the degree of overlap of the lines becomes:

$$\overline{\beta_{L/D}^*}=\frac{1}{\overline{k^*u}}\int_0^l x(s)\,p(s)\,\overline{k}(s)\,\overline{\beta_{L/D}}\,ds \qquad [5.29]$$

5.1.6.3. *Box model*

In the box model, we neglect the spectral correlations, causing the parameter β to disappear. We are left with only the parameter $\overline{k}$, which is the reduced absorption coefficient, or indeed the “box model” parameter.

With the integral form of the RTE averaged over $\Delta\nu$ for a uniform column of length l, we find the following with a good degree of approximation:

$$\overline{\tau}^{\Delta\nu}(l) = e^{-x p \overline{k} l} \qquad [5.30]$$

For a non-uniform column, we obtain:

$$\overline{\tau}^{\Delta\nu}(l) = e^{-\int_0^l x(s) p(s) \overline{k}(s) ds} \qquad [5.31]$$

5.1.6.4. *Other models*

It should also be mentioned that aside from the band models used by ONERA, it is also possible to use more detailed models, termed "ray-by-ray", such as that of the PARADE code (*Plasma Radiation Database*) run by ESA [DUB 06, RIT 13] or SPARTAN (*Space-Related Heat Transfer Analysis*) [DAS 04], coupled with the so-called *ray tracing* methods [DAS 13] used at the IST in Lisbon, to determine the radiative fluxes on a wall, taking account of the emission and absorption.

5.1.7. *Modeling of the radiative properties of the particles*

5.1.7.1. *Expressions of the radiative properties*

The intensity is modified when interaction occurs between the photons and the particles of a multi-phase medium (internal combustion engines with soot or particles of aluminum oxide) by the effect of absorption and scattering of the radiative transfer.

We can characterize the radiative properties of a cloud of spherical particles of radius r interacting with an incident radiative transfer of wavenumber ν, by the following two dimensionless parameters [MOD 03]:

– the complex optical index of the particles: $m(s,\nu) = n(s,\nu) - ik(s,\nu)$, a function of the wavenumber and the thermo-physical conditions at s;

– the size parameter (or Mie parameter): $x = 2\pi r \nu$.

The general theory, which was first advanced by Mie, describes the interaction between an electromagnetic wave and a spherical particle. There are two particular cases which limit the applicability of Mie's theory:

– $x << 1$, this is the Rayleigh theory;

– $x >> 1$, this is the domain of geometrical optics.

If $x = O(1)$, we need to consider the complete Mie theory.

The effectiveness factors Q_{abs} and Q_{sca} quantify the importance of each phenomenon. They are the ratios between the effective sections of absorption and diffusion of the radiative transfer. Figure 5.2 shows an example with aluminum oxide particles [TES 11].

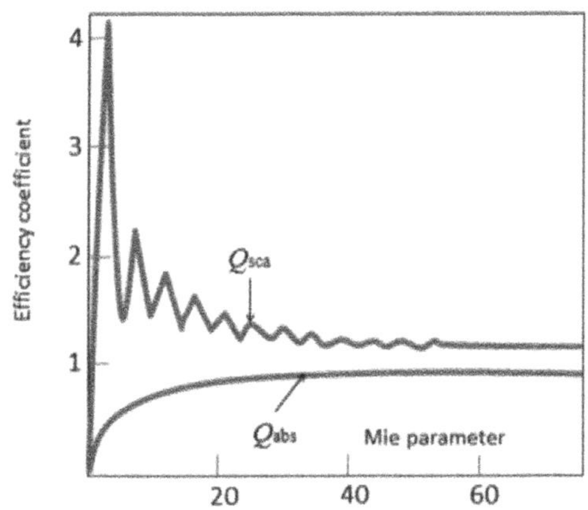

Figure 5.2. *Effectiveness coefficients for absorption and scattering of the radiative transfer by aluminum oxide particles, for a complex refraction index equal to* $m = 1.7 - 2\,10^{-3}\,i$ *(figure redrawn from [TES 11], with the authors' kind permission)*

The monochromatic absorption $\kappa_\nu(s)$ and scattering coefficients $\sigma_\nu(s)$ in terms of s are determined by means of the effectiveness coefficients and the volume particle size distribution (PSD) $f_\nu(r,s)$:

$$\begin{cases} \kappa_\nu(s) = \dfrac{3}{4}\displaystyle\int_0^\infty \dfrac{Q_{abs}\left[m(s),r,\nu\right]}{r} f(r,s)\,dr \\ \sigma_\nu(s) = \dfrac{3}{4}\displaystyle\int_0^\infty \dfrac{Q_{sca}\left[m(s),r,\nu\right]}{r} f(r,s)\,dr \end{cases} \quad [5.32]$$

To characterize anisotropic radiative scattering, we also need the phase function $\Phi_\nu(s,\mathbf{u}'\rightarrow\mathbf{u})$, which is a weighted integral of the sizes, the phase function $\Phi_\nu(s,\mathbf{u}'\rightarrow\mathbf{u},r)$ given by Mie's theory for each spherical particle size [TES 11, MOD 03].

5.1.7.2. *Box model for particles*

For simplicity's sake, it is useful to model the particles in a suspension with the same band spectral discretization $\Delta\nu$ as is used for the gaseous phase. Given that, in general, the dynamic of the particles' absorption spectrum is sufficiently low to be able to discount spectral correlations, we can use the box model. We therefore write the average absorption coefficient for a band:

$$\bar{\kappa}^{\Delta\nu}(s) = \sum_i \frac{1}{\Delta\nu}\frac{3}{4}\int_{\Delta\nu,\Delta r_i} \frac{Q_{abs}\left[m(s),r,\nu\right]}{r} f(r,s)\,dr\,d\nu \quad [5.33]$$

with Δr_i representing discretization in terms of particle radii.

If the spectral bands are sufficiently narrow to consider the complex index m to be constant in each band, the above integration will relate only to the Mie parameter x, with a distribution function $g(x-x_i)$ describing the variation of x

due to the PSD within the range Δr_i and the variations of the wavenumber within the range $\Delta \nu$ [JOU 01].

5.1.7.3. *Rayleigh regime for scattering*

As indicated in section 5.1.7.1, if the particle sizes are sufficiently small in relation to the wavelength of the light, Mie's parameter is such that $x << 1$, and we are witnessing a Rayleigh scattering regime [MOD 03].

In this regime, the effectiveness factors are such that $Q_{abs} \propto x$, $Q_{sca} \propto x^4$, so the effect of the scattering of light by the particles is negligible in relation to that of absorption. Therefore, only one of these coefficients plays a part: the absorption coefficient, which is written as [MOD 03]:

$$\kappa_\nu(s) = A(s,\nu)\nu f_V(s) \qquad [5.34]$$

where:

$$A(s,\nu) = 36\pi n(s,\nu)k(s,\nu) \Big/ \left[\left(n^2(s,\nu) - k^2(s,\nu) + 2\right)^2 + \left(2n(s,\nu)k(s,\nu)\right)^2\right]$$

and $f_V(s)$ is the local volume fraction of particles. $A(s,\nu)$ depends explicitly on the real and imaginary parts of the local complex optical index. For a sufficiently narrow spectral band, we can write an average absorption coefficient:

$$\overline{\kappa}^{\Delta\nu}(s) = A(s,\nu_c)\nu_c f_V(s)$$

5.2. Application to the hypersonic flow of atmospheric re-entry

The problem of re-entry of a body (a satellite, a spacecraft, a missile, a meteorite) into the atmosphere is multi-faceted. Discounting the scenario where we wish to destroy that body, it is of the greatest possible importance to know the temperatures that are reached and therefore the radiative fluxes experienced by the wall of the object.

It is important to mention the efforts that have been made to experimentally simulate the conditions of re-entry to planetary atmospheres [KAM 13]. However, for our purposes here, we shall limit ourselves to the theoretical and numerical modeling efforts.

Below, we give two examples of how to deal with this problem: one greatly simplified, because we merely examine the one-dimensional flow on the stagnation line between the detached shockwave and the surface of the object (see Figure 5.3), using the system of equations [5.19]; the other, which is more complex and three-dimensional, is detailed in Lamet's thesis [LAM 09].

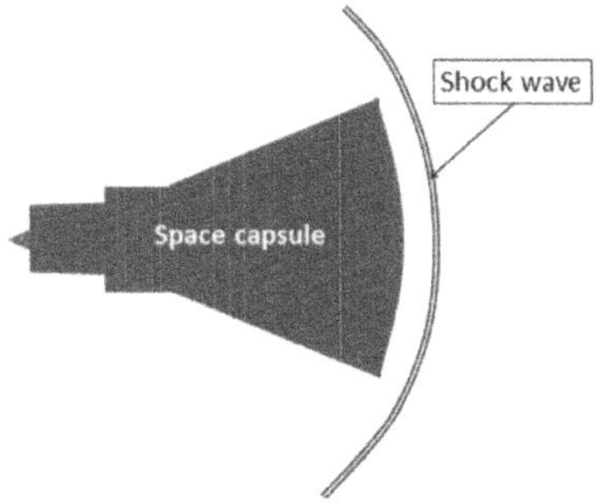

Figure 5.3. *Detached shockwave traveling in front of a body undergoing atmospheric re-entry (drawn from a photo released by NASA about the Mercury Program, 1957)*

5.2.1. *One-dimensional approximation for a re-entering body*

We have examined many cases of one-dimensional flows in the first two volumes of this series, and in the early chapters of this volume.

For high-velocity re-entry of satellites into the atmosphere, we can adopt the one-dimensional approximation for the part of the flow lying between the shockwave and the stagnation point situated at the nose of the re-entering body.

A number of authors have been working on this problem since the 1960s [PAR 85, PAR 89].

In a recent article by Allouche *et al.* [ALL 13], the physio-chemical processes encountered in hypersonic flows, such as vibratory excitation, dissociation and ionization at high temperatures behind a powerful detached shockwave, are modeled by using the vibration relaxation model with CVD coupling and the equations given in section 5.1.4.1, governing the phenomena of radiative transfer. A complete calculation of the flow is performed behind the shockwave, in the conditions of a body experiencing re-entry into the Earth's atmosphere: 10,000 m/s, 10 Pa, 300 K. This calculation, by finite differences, integrates the radiative transfer and the various forms of disequilibrium (chemical, *between the internal modes of vibration* and translation/rotation) and couplings between dissociation and vibration, with the 11 chemical species considered by Park [PAR 85]: O, N, NO, O_2, N_2, O^+, N^+, NO^+, O_2^+, N_2^+, e^- and 47 chemical reactions.

We proceed by successive iterations on the discretized equations. A convergence test is performed on the temperature, before a re-evaluation of the characteristic values for each node in the relaxation range, until equilibrium is reached.

The model used integrates four temperatures (T_{tr}, T_{vO2}, T_{vNO} and $T_{vN2} = T_e$) characterizing the modes of energy of translation-rotation of the neutral particles and ions, the molecular vibrations and the translation of the electrons. This model performs better than the two-temperature model used by other authors, because it describes the post-shock boundary layer and the diffusive nature of the vibration relaxation at high temperature (see [ALL 13] and also section 1.3 of [PRU 13], along with [FLA 90], in relation to flows at thermal and chemical disequilibrium).

Furthermore, the chemical mechanism with 47 reactions used here, with a shockwave speed of 10 km/s, corresponds more closely to the report published by NASA [GUP 90], which defines the regime of flow of capsules during their re-entry into the atmosphere.

The temperature T, which is initially 49,000 K behind the shockwave, decreases and reaches 9,900 K at equilibrium and coincides with T_v (in Park's work [PAR 89], it decreased steadily before stabilizing at 13,152 K). This significant reduction is obtained because of a reactional mechanism containing a greater number of endothermic reactions, such as the formation of species by dissociation reactions, and numerous ions being formed by ionization reactions. Also, the effect of radiative transfer, of which account is taken in this work, contributes to the reduction in temperature of the flow.

Compared to the previous results, the results of Allouche *et al.*'s work [ALL 13] give a lower temperature of impact on the wall of space vehicles (resulting from the rapid reduction in temperature of the flow after the shock and the extent of the zone of relaxation to equilibrium).

5.2.2. *3D calculations for a body experiencing re-entry*

Figure 5.4 shows an example of the result of calculations obtained by J.-M. Lamet [LAM 09]. The results were obtained using the Monte Carlo method [DUP 06, MEH 10] with the ASTRE code used by ONERA [TES 01].[4]

At ONERA, the Monte Carlo method (the ASTRE solver) and the discrete ordinate method (REA solver) are included in the CEDRE code[5] [REF 11].

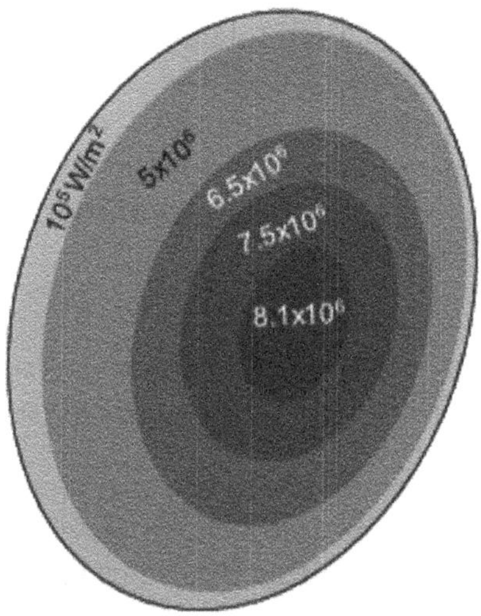

Figure 5.4. *Radiative fluxes at the wall for 185.6 million rays. Trajectory point 1643s from the FIRE II probe. Extract modified from Jean-Michel Lamet's thesis [LAM 09], with the author's permission*

These methods are described in [TES 11] where, in the interests of clarity of the discussion, the expressions used are

4 The *ASTRE* code (which stands for *Approche statistique des transferts radiatifs dans les écoulements* – Statistical approach to radiative transfers in flows) is a 3D radiative transfer code based on a Monte Carlo approach. Developed in 1998 to study turbulent flames, its applications have extended to cases of Martian atmosphere re-entries for media at LTE (Soubrié's thesis in 2006 [SOU 06]).

5 The *CEDRE* code (standing for *Calcul d'écoulements diphasiques et réactifs pour l'énergétique* – Computation of two-phase reactive flows for energy applications) was developed at ONERA (see section 4.1.5).

monochromatic, whereas in reality these methods are used with band spectrum models.

Finally, it should be mentioned that a great deal of activity surrounding bodies undergoing atmospheric re-entry has been being coordinated, for many years now, by the ESA, with the participation of a dozen European labs and institutions [RIT 13]. Technological roadmaps have been defined, and are regularly updated, in relation to planetary explorations, including – amongst other aspects – the problems of radiative transfers during atmospheric re-entry, protection of the astronauts from radiation, and the study of ablative materials.

5.3. Application to the boundary layer above a flat plate with soot formation and radiative transfer

Fires involve the combustion of solid bases in the presence of air flows. On the ground, these flows are often due to buoyancy forces. On satellites in a microgravity environment, other flows are present, but the behavior of the flames is different. In particular, it was the lack of existence of fire regulations and procedures on board spacecraft which led the LCD at the CNRS in Poitiers, headed by Pierre Joulain, to develop fundamental research in this domain, and persuaded the CNES and the ESA to support that work [JOU 93, VIE 99, WAN 01].

Before attacking the main theme of this section, it is useful to recap on a few classic results relating to boundary layers in multi-component fluids: boundary layers with diffusion and the Emmons problem.

The notion of a viscous boundary layer was examined, in the case of a single-component fluid, in section 1.7.2.3 of [PRU 12]. Presentations of the non-reactive problem by means of the asymptotic expansion method have been given by numerous authors.

5.3.1. *Recap on a boundary layer with diffusion*

Let us imagine that diffusion does not modify the flow velocity profile. If it is a Blasius profile, we have:

$$\begin{cases} u = U_\infty f'(\eta),\ \eta = y\sqrt{U_\infty/2\nu x}, \\ v = \sqrt{\nu U_\infty/2x}\left[\eta f'(\eta) - f(\eta)\right], \\ f''' + f f'' = 0,\ f'(0) = f(0) = 0,\ f'(\infty) = 1 \end{cases} \qquad [5.35]$$

We can distinguish numerous cases, which depend on the values of the Schmidt number $Sc = \nu/D$.

If Sc>>1: suppose, to begin with, that diffusion takes place very near to the wall and that the thickness of the diffusion boundary layer is much less than that of the viscous boundary layer: $\delta_C \ll \delta$. In that zone, we would have:

$$u \cong U_\infty \eta f''(0) \qquad [5.36]$$

The balance of species equation is written thus:

$$u\partial Y/\partial x + v\partial Y/\partial y - D\partial^2 Y/\partial y^2 = 0 \qquad [5.37]$$

in the case of a Fick law with a single diffusion coefficient, and if $\partial^2 Y/\partial x^2 = O(1/x^2)$ is negligible in comparison to $\partial^2 Y/\partial y^2 = O(1/\delta_C^2)$.

The first and last term are of the same order of magnitude, so that $u/x \approx D/\delta_C^2$, or indeed: $U_\infty f''(0)\sqrt{U_\infty/2\nu x}\,\delta_C/x \approx D/\delta_C^2$.

Because we have $\delta \approx \sqrt{2\nu x/U_\infty}$, we can deduce:

$$\delta_C/\delta \approx \left[D/2\nu f''(0)\right]^{1/3} = O\left(Sc^{-1/3}\right) \qquad [5.38]$$

As the Schmidt number is equal to $Sc = \nu/D$, we can see that this case does indeed correspond to Sc>> 1.

Suppose now that a very fast chemical reaction $A \rightarrow B$ takes place, corresponding to catalysis at the wall. At $y = 0, (\eta = 0)$, we have: $Y_A = 0, Y_B = 1$, and at $y = \infty(\eta = \infty)$: $Y_A = 1, Y_B = 0$. In the vicinity of the wall, at $y = O(\delta_C)$:

$$u = U_\infty f''(0)\eta,\ v = \sqrt{U_\infty \nu/2x} f''(0)\eta^2/2 \qquad [5.39]$$

Thus, $Y_B = Y(\eta)$. We obtain:

$$d^2Y/d\eta^2 + \left(\nu f''(0)\eta^2/2D\right)dY/d\eta = 0 \qquad [5.40]$$

The solution to this equation is:

$$Y = 1 - I(\eta)/I(\infty) \qquad [5.41]$$

where: $I(\eta) = \int_0^\eta \exp\left(-f''(0)Sc\,\xi^3/6\right)d\xi$. Of course, here we see again the same order of magnitude as the thickness of the boundary layer of the concentrations: $\delta_C/\delta = O\left(Sc^{-1/3}\right)$.

If Sc<<1: let us now suppose that the Schmidt number is small. The phenomenon of diffusion takes place in a layer whose thickness is $\delta_C >> \delta$, and the concentrations vary very little within the viscous boundary layer, so that almost everywhere, we have $u = U_\infty$. The diffusion equation becomes:

$$U_\infty \partial Y/\partial x - D\partial^2 Y/\partial y^2 = 0 \qquad [5.42]$$

and the thickness δ_C is therefore such that: $U_\infty/x \approx D/\delta_C{}^2$, which gives us:

$$\delta_C/\delta \approx \sqrt{D/2\nu} = O\left(Sc^{-1/2}\right), Sc << 1$$

In the case of the reaction examined above, we therefore obtain the solution:

$$Y = 1 - \operatorname{erf}\left(\sqrt{\mathrm{Sc}/2}\,\eta\right) \quad [5.43]$$

– *If Sc=1:* finally, if the Schmidt number is equal to 1, we have:

$$\begin{cases} u\,\partial u/\partial x + v\,\partial u/\partial y - \nu\,\partial^2 u/\partial y^2 = 0 \\ u\,\partial Y/\partial x + v\,\partial Y/\partial y - D\,\partial^2 Y/\partial y^2 = 0 \end{cases} \quad [5.44]$$

where $D = \nu$.

The viscous and diffusion boundary layers are of the same thickness, and in the particular case of a rapid reaction at the wall, we therefore find the self-similar solution:

$$u/U_\infty = 1 - Y = f'(\eta) \quad [5.45]$$

where $f(\eta)$ satisfies the Blasius equation with: $f'(0) = f(0) = 0, f'(\infty) = 1$.

– If *Sc~1:* if Sc is around 1 but not exactly equal to it, it is less easy to find the solution. In such cases, we need to solve the diffusion equation, if we know the velocity profile.

Finally, here, for illustrative purposes, we have only looked at situations where the chemical process is controlled by diffusion. In reality, very often there is an interaction between diffusion, the chemical kinetics and heat transfer. The example discussed below illustrates such interactions.

5.3.2. *Reminders about the Emmons problem*

A flat ablative plate (see Figure 5.5) is sublimated with an amount of heat ΔL in a gaseous stream of oxidizing agent. The fuel plate has a density ρ_s, and by its sublimation, releases H gas. Upon contact with the oxidant O, the following exothermic chemical reaction occurs: $H + O \rightarrow P$. The heat involved in this reaction in standard conditions is:

$$\Delta H = -\sum_j \nu_j M_j \left(q_f^0\right)_j \qquad [5.46]$$

The phenomenon includes heat conduction, diffusion of the species present, and viscosity. We shall accept the hypothesis of a negligible pressure gradient, and assume that the Schvab–Zeldovich approximation [PRU 12] can be applied. The flow is two-dimensional plane and steady. We can easily demonstrate that the terms in $\partial^2/\partial x^2$ are negligible in comparison to the terms in $\partial^2/\partial y^2$ because the layer is very thin: $\delta(x) << x$ for a sufficiently high value of x (here, we are sufficiently far away from the leading edge, but the flow nevertheless remains laminar).

The equations are as follows for a flow of compressible, reactive fluid, using the notations set out in section 2.5.2 of [PRU 12]:

$$\begin{cases} \partial(\rho u)/\partial x + \partial(\rho v)/\partial y = 0 \\ \rho u \partial u/\partial x + \rho v \partial u/\partial y - (\partial/\partial y)\left(\mu \partial u/\partial y\right) = 0 \\ \rho u \partial\beta_j/\partial x + \rho v \partial\beta_j/\partial y - (\partial/\partial y)\left(\rho D \partial\beta_j/\partial y\right) = \dot{\zeta} \\ \rho u \partial\beta_T/\partial x + \rho v \partial\beta_T/\partial y - (\partial/\partial y)\left[\left(\lambda/c_{p,f}\right)\partial\beta_T/\partial y\right] = \dot{\zeta} \end{cases} \qquad [5.47]$$

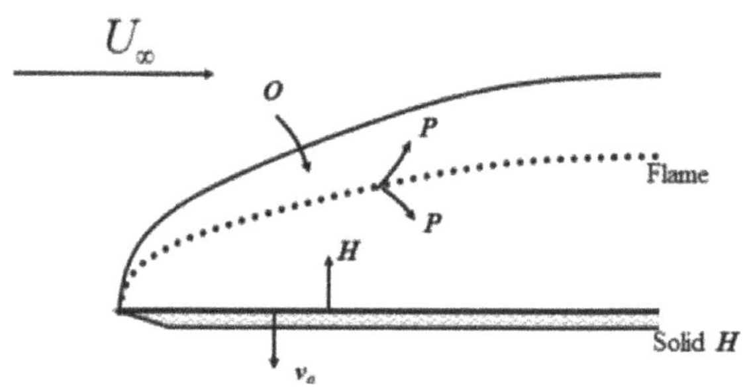

Figure 5.5. *The Emmons problem with a slender flame*

The wall conditions are obtained using the balance equation at the discontinuity (Chapter 3 of [PRU 12]):

$$\left[\mathcal{J}_F + \rho f\left(\mathbf{v} - \mathbf{V}_S\right)\right]_-^+ \cdot \mathbf{N} = 0 \qquad [5.48]$$

with the (–) side corresponding to the solid and the (+) side to the fluid. When applied to the mass, the momentum, the chemical species (no surface reaction takes place) and the total energy, relation [5.48] gives us the system:

$$\begin{cases} (\rho v)_P = \rho_s v_a \\ \tau_{xy} = (\mu \partial u/\partial y)_P \\ -(\rho D \partial \beta_j/\partial y)_P + (\rho v)_P \beta_{jP} = \rho_s v_a \beta_{js} \\ -\left[(\lambda/c_{p,f})\partial \beta_T/\partial y\right]_P + (\rho v)_P \beta_{TP} = -\rho_s v_a l/\Delta H \end{cases} \qquad [5.49]$$

with v_a being the ablation rate, τ_{xy} the unknown friction tension, and s the solid phase supposed to be of uniform density and chemical composition.

The gaseous medium is compressible. This difficulty is easily overcome by *performing the Howarth variable change*:

$$\xi = \int_0^x \rho\mu \, dx \qquad [5.50]$$

which gives us $d\xi = \rho\mu\, dx$, if we suppose that the value $\rho\mu$ depends solely on the abscissa *x*:

$$\zeta = \int_0^y \rho\, dy \Leftrightarrow d\zeta = \left[\int_0^y (\partial\rho/\partial x)\, dy \right] dx + \rho\, dy \qquad [5.51]$$

Finally, we set:

$$g = (1/\rho u)\left[\rho v + u \int_0^y (\partial\rho/\partial x)\, dy \right] \qquad [5.52]$$

The resulting system of equations is therefore:

$$\begin{cases} \partial u/\partial\xi + \partial g/\partial\zeta = 0, \quad d\psi = u\, d\zeta - g\, d\xi \\ u\,\partial u/\partial\xi + g\,\partial u/\partial\zeta - \partial^2 u/\partial\zeta^2 = 0 \\ u\,\partial\beta_j/\partial\xi + g\,\partial\beta_j/\partial\zeta - (\partial/\partial\zeta)\left(Sc^{-1}\,\partial\beta_j/\partial\zeta\right) = \dot{\zeta}/\rho^2\mu \\ u\,\partial\beta_T/\partial\psi + g\,\partial\beta_T/\partial\zeta - (\partial/\partial\zeta)\left(\Pr^{-1}\,\partial\beta_T/\partial\zeta\right) = \dot{\zeta}/\rho^2\mu \end{cases} \qquad [5.53]$$

with, at the wall:

$$\begin{cases} (\rho\mu g)_p = \rho_s v_a \\ \tau_{xy} = \rho\mu(\partial u/\partial\zeta)_p \\ -Sc^{-1}\left(\partial\beta_j/\partial\zeta\right)_p + g\left(\beta_j - \beta_{js}\right)_p = 0 \\ -\Pr^{-1}\left(\partial\beta_T/\partial\zeta\right)_p + g\,l/\Delta H = 0 \end{cases} \qquad [5.54]$$

The first two equations in terms of u and g in system [5.53] are identical to those of the incompressible boundary layer, and the system is therefore able to have self-similar solutions. The thickness of that boundary layer is, in direction ζ : $\delta(\xi) = \sqrt{\xi/U_\infty}$ and we set:

$$\eta = \zeta\sqrt{U_\infty/2\xi} \qquad [5.55]$$

The velocity has the components:

$$u = U_\infty f'(\eta), \quad g = \sqrt{U_\infty / 2\xi}\left[\eta f'(\eta) - f(\eta)\right] \qquad [5.56]$$

and the stream function becomes:

$$\psi = \sqrt{2\xi U_\infty f(\eta)} \qquad [5.57]$$

The wall conditions give us:

$$\begin{cases} -\rho\mu\sqrt{U_\infty / 2\xi}\, f(0) = \rho_s v_a \\ \tau_{xy} = \rho_P \mu_P \sqrt{U_\infty / 2\xi}\, f''(0) \end{cases} \qquad [5.58]$$

We can see that the ablation rate is in $\xi^{-1/2}$. The value of $f(0)$ will be determined by the other conditions relating to β_j and β_T. Additionally, we have:

$$f'(\infty) = 1, f'(0) = 0 \qquad [5.59]$$

When the Schmidt and Prandtl numbers are both equal to 1, the parameters $(\beta_j - \beta_T)$ satisfy the same equation as u. One particular solution to the problem is the *linear Crocco relation*:

$$\beta_T - \beta_O = (\beta_{T\infty} - \beta_{TP} - \beta_{O\infty})u/U_\infty + \beta_{TP} \qquad [5.60]$$

This is the solution to the problem, as we can show that the solution is unique.

At the wall, we have:

$$\partial\beta_T / \partial\zeta|_P = g(0) l / \Delta H = -\sqrt{U_\infty / 2\xi}\, f(0) l / \Delta H$$
so: $$\partial\beta_T / \partial\eta|_P = -f(0) l / \Delta H \qquad [5.61]$$

The Crocco relation also gives us:

$$\partial\beta_T/\partial\eta\big|_P = (\beta_{T\infty} - \beta_{TP} - \beta_{O\infty})\, f''(0) \qquad [5.62]$$

because $\beta_O\big|_P = 0$, if we suppose that all the oxidant is consumed in the flame.

Let us now introduce the *Spalding parameter*:

$$B = (\beta_{T\infty} - \beta_{TP} - \beta_{O\infty})\,\Delta H/l \qquad [5.63]$$

The three boundary conditions:

$$f'(\infty) = 1,\, f'(0) = 0,\ f(0) = -B\, f''(0) \qquad [5.64]$$

enable us to solve the Blasius equation $f''' + f\, f'' = 0$.

Indeed, the value B gives us $f(0)$ (Figure 5.6), and therefore the rate of regression of the fuel:

$$v_a = \sqrt{U_\infty/2\xi}\; f(0)\, \rho\mu\big|_P \big/ \rho_H \qquad [5.65]$$

where $\rho_H = \rho_s$ is the density of the solid fuel, the friction tension on the plate:

$$\tau_{xy} = U_\infty \sqrt{U_\infty/2\xi}\; f''(0)\, \rho\mu\big|_P \qquad [5.66]$$

and the temperature gradient at the wall:

$$\left.\frac{\partial\beta_T}{\partial\eta}\right|_P = -f(0)\frac{l}{\Delta H} \qquad [5.67]$$

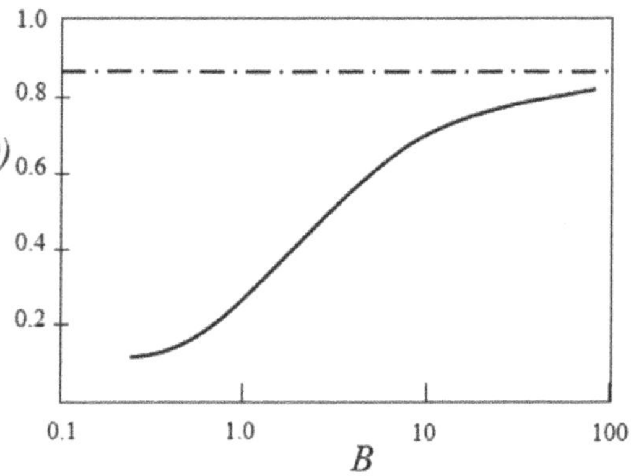

Figure 5.6. *The value $(-f)$ at the wall as a function of the Spalding parameter*

These results are independent of the chemical kinetics and of the position of the flame.

If the flame is taken to be slender, the rate of chemical production $\dot{\zeta}$ is zero outside of the value η_f (which gives us a parabola $\zeta^2 = \eta_F{}^2 2\xi/U_\infty$), where the stoichiometric ratio is $Y_O/Y_H\big|_f = r_{st}$. Thus, we have the following concentration profiles:

$$\begin{cases} Y_O = Y_{O\infty}\left[f'(\eta_f)-f'(\eta)\right]/\left[f'(\eta_f)-1\right],\ Y_H = 0, \text{ for } \eta > \eta_f, \\ Y_O = 0,\ Y_H = f(0)\left[f'(\eta_f)-f'(\eta)\right]/\left[f(0)f'(\eta_f)-f''(0)\right], \\ \text{for } 0 < \eta < \eta_f \end{cases} \quad [5.68]$$

The value is η_f is determined by the equation $Y_H\big|_f = r_{st}\,Y_O\big|_f$, so:

$$r_{st}Y_{O\infty} = f(0)\left[f'(\eta_f)-1\right]/\left[f(0)f'(\eta_f)-f''(0)\right] \quad [5.69]$$

The profiles for the concentrations and for the temperatures can be deduced from this. The local friction coefficient is:

$$C_F = \frac{\tau_{xy}}{\rho U_\infty^2/2} = \left(\frac{2\mu_\infty^2}{\xi U_\infty}\right)^{1/2} f''(0) = \frac{2}{B}\frac{\rho_s v_a}{\rho_\infty U_\infty} \qquad [5.70]$$

The present theory adopting the Shvab–Zeldovich hypotheses requires the absence of gravity. In the presence of gravity, the results are very different indeed.

The results of experiments carried out in micro-gravity conditions exhibit certain flame profiles corresponding to Emmons' theory. However, this is not generally the case.

Among the important factors neglected by this theory, it is noteworthy that it does not take account of the soot which is always present in the case of combustion of hydrocarbons and systems generating energy losses by radiative transfer. Energy is also lost to the wall.

5.3.3. *Influence of soot and radiative transfer*

Torero *et al.* [TOR 02], who study the flames of accidental fire aboard spacecraft, proposed a *mass transfer parameter* in the absence of gravity, which is different to the Spalding parameter but exhibits close correspondence with experiments carried out in microgravity. That number is as follows[6]:

$$B = \frac{(1-\chi)\Delta H_c Y_{O_2\infty} + c_{p\infty}(T_f - T_\infty)}{\Delta H_p + Q} \qquad [5.71]$$

6 In Torero *et al.* [TOR 01], the flame temperature T_f in the numerator is replaced by the wall temperature T_p.

where χ is the radiative loss to the surrounding environment between the flame and the surface; ΔH_c is the enthalpy of combustion per unit mass of oxygen consumed; $Y_{O_2\infty}$ is the mass fraction of oxygen in ambient conditions; $c_{p\infty}$ is the specific heat at constant pressure in ambient conditions; T_f is the temperature of the flame; T_∞ the ambient temperature of the mixture; ΔH_p is the enthalpy of pyrolysis per unit mass of fuel pyrolyzed; and Q is the total loss of enthalpy at the surface per unit mass of pyrolyzed.

This loss includes heat conduction through the ablative plate, re-emission from the surface by ablation (which can be estimated on the basis of the temperature of pyrolysis), and radiative reflux stemming from the flame.

In theory, we can evaluate χ and Q by analyzing the simple juxtaposition of Emmons' solution against the experimental profile. Rangwala [RAN 08] sought a way of improving the correspondence between theory and reality by using a variable parameter B.

The modeling of diffusion flames is the subject of a great many theoretical and numerical studies (see Figure 5.7), and experimental studies [FER 84, LIG 07].

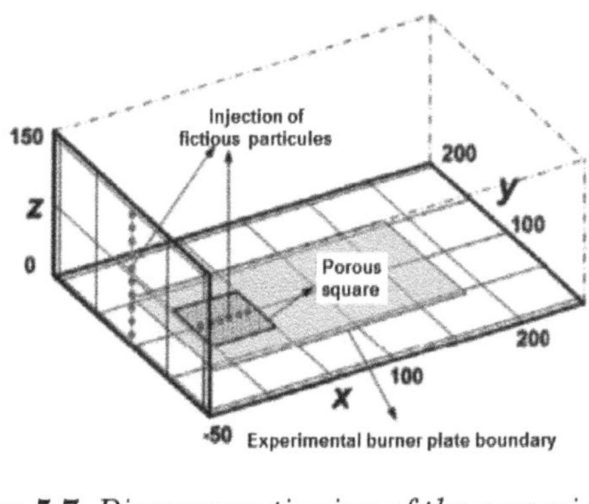

Figure 5.7. *Diagrammatic view of the numerical domain (G. Legros, personal communication, 2014; also see [LEG 09])*

One of the most difficult problems to solve is that of the model of the interaction of soot with the rest of the flow [WAN 07, DEM 12]. A soot formation mechanism was put forward by Bockhorn in 1994 [BOC 94]. That mechanism comprises the following consecutive steps: a molecular arena containing fuel, oxidant and reaction products (characteristic entity size: 0.5 nm), a platform of particles which can be subdivided at the start of the process, increase of the surface area and coagulation (characteristic entity size: 50 nm). This question of soot formation was also touched upon in Chapter 1 of this book.

In order to simplify the experimental setup, we often use gaseous fuel injected by a porous square [LEG 06], and the validity of the two-dimensional hypothesis can be evaluated experimentally [LEG 05].

More recent studies on soot (both numerical and experimental) have been published by Legros *et al.* In [LEG 09], the authors integrated new numerical models with the pre-existing ones and the experimental observations, in order to provide a coherent explanation for the lengths of the flame and the volume fractions of soot observed with laminar diffusion flames.

5.4. Application to combustion of aluminum-based solid propellants

In this section, we give the fundamental information about "black powder rockets". The powders in question are solid propellants, which are homogenized mixtures of fuels and solid oxidizing agents, designed to burn in the absence of any external supply. The bricks of powder are generally cylindrical and hollow along the axis, with an interior surface that is cylindrical and has a more or less complicated straight section – in principle to facilitate ignition and combustion. These bricks of powder are ground with extreme caution so as to prevent any accidents.

During the combustion process, the aluminum particles scraped off the powder brick become liquefied, and the drops which result from that liquefaction evaporate. The combustion of aluminum vapor with oxygen from the solid oxidizing agent produces aluminum oxide in the form of a solid shell surrounding the drop, which disintegrates more or less quickly into fine particles.

The condensed particles, be they liquid or solid, raised to high temperatures, emit radiative transfer which needs to be taken into account by any model.

A review of knowledge in the physics and modeling of the combustion of aluminum-based droplets was compiled by Dupays *et al.* [DUP 00]. Particular mention is made of the numerical efforts under way at ONERA with the MSD (3D Multi-Species Navier–Stokes code). The following mechanism including 9 chemical reactions, established by Catoire (CNRS/LCSR in Orléans), yields very good results:

$$
\begin{array}{llllllll}
Al+O_2 & \Leftrightarrow & AlO+O & (1) & AlO & \Leftrightarrow & Al+O & (2) \\
AlO+O_2 & \Leftrightarrow & AlO_2+O & (3) & AlO_2 & \Leftrightarrow & AlO+O & (4) \\
Al_2O & \Leftrightarrow & Al+AlO & (5) & Al_2O_2 & \Leftrightarrow & AlO+AlO & (6) \\
Al_2O_2 & \Leftrightarrow & AlO_2+Al & (7) & Al_2O_2 & \Leftrightarrow & Al_2O+O & (8) \\
O_2+M & \Leftrightarrow & O+O+M & (9) & & & &
\end{array}
\quad [5.72]
$$

One of the problems encountered with solid aluminum-based propellants is the accumulation of very significant masses of scoria of aluminum oxide in certain areas of the engine, upstream of the output nozzle [TOT 05]. Modeling efforts have been made, taking account of the turbulence by using LES methods. Work is continuing with a view to revealing the recirculation eddies which are responsible for these deposits, and finding solutions to eliminate them – for instance, by installing water sprays in the appropriate places.

Tessé and Lamet [TES 11] give the following summary of the method used at ONERA to take account of the effects of radiative transfer.

In the combustion chamber of a solid propellant engine, temperatures of up to 3,600 K can be attained, and a pressure of between 50 and 150 bars. At high pressure, because of the effect of expansion due to the collisions, spectral dynamics show less marked absorption spectra than those found at atmospheric pressure.

In this situation, the radiative transfer in the chamber can be calculated without taking account of the spectral correlations. Thus, we can model the radiative properties of the gases using the "box model", developed by Duval [DUV 02], by taking account of the species CO_2, H_2O, CO and HCl. The parameters $\bar{k}$ are tabulated for a range of temperatures between 300 and 3,500 K, and for 43 spectral bands of varying width variable between 150 and 9,300 cm^{-1}.

The radiative properties of aluminum oxide particles are calculated with Mie's theory. Using a distribution function $g(x-x_i)$, as indicated in section 5.1.7.2, we write:

$$\bar{\kappa}^{\Delta\nu}(s)=\sum_i \frac{3}{4} f_{V,i}(s)\int_{\Delta x_i}\frac{Q_{abs}\left[m(s),\nu\right]}{x}g(x-x_i)\,dx \qquad [5.73]$$

We obtain a similar expression for the scattering $\bar{\sigma}^{\Delta\nu}(s)$:

$$\bar{\sigma}^{\Delta\nu}(s)=\sum_i \frac{3}{4} f_{V,i}(s)\int_{\Delta x_i}\frac{Q_{sca}\left[m(s),\nu\right]}{x}g(x-x_i)\,dx \qquad [5.74]$$

For the phase function $\Phi_\nu(s,\mathbf{u}'\rightarrow\mathbf{u})$, we also obtain:

$$\bar{\Phi}^{\Delta\nu}(s,\mathbf{u}'\rightarrow\mathbf{u})=\sum_i \frac{3}{4}\frac{f_{V,i}(s)}{\bar{\sigma}^{\Delta\nu}(s)}\int_{\Delta x_i}\frac{Q_{sca}\left[m(s),\nu\right]}{x}\Phi_x(s,\mathbf{u}'\rightarrow\mathbf{u})g(x-x_i)dx \qquad [5.75]$$

The absorption and scattering coefficients and phase functions are calculated with the hypothesis whereby the distribution function g in equations [5.72], [5.73] and [5.74] is a Gaussian. In general, several classes of particle sizes are used (typically ranging between one and five) to describe the particle size distribution. The complex refraction index of the aluminum oxide is modeled as a function of the wavenumber and of the temperature in conformity with the expressions given by Dombrowski [DOM 82] or Joumani [JOU 01].

Appendix

Concepts Surrounding the Hopf Bifurcation

The earliest concepts regarding bifurcations were given in Appendix A5 of [PRU 13]. There, we used these concepts in the problem of change of stability of the flow between two cylinders in differential rotation. For this purpose, we limited ourselves to supercritical *pitchfork* bifurcation.

This time, in Chapter 1 of this volume, we are looking at the analysis of the influence of an acoustic wave on a fog flame [NIC 09], which involves the concept of the *Hopf bifurcation*.[1]

Let us begin with an example: that of the following nonlinear dynamic system, expressed in Cartesian coordinates in the plane (x, y):

$$\begin{cases} dx/dt = \mu x - y - x\left(x^2 + y^2\right) \\ dy/dt = \mu y + x - y\left(x^2 + y^2\right) \end{cases} \qquad [A.1]$$

1 The instability presented here was studied, in its time, by Poincaré and also by Andronov and by Hopf.

This system is expressed as follows in polar coordinates:

$$\frac{dr}{dt} = r\left(\mu - r^2\right); \quad \frac{d\theta}{dt} = 1 \qquad [A.2]$$

If the parameter μ is non-null and constant, we obtain the following, with A and B being integration constants:

$$\begin{cases} \theta = t + A \\ \left|1 - \mu/r^2\right| = B e^{-2\mu t} \end{cases} \qquad [A.3]$$

The constant A can be taken to be equal to zero. If we use the notation r_0 for the initial value of r, we have: $B = \left|1 - \mu/r_0^2\right|$ and the system [A.3] becomes:

$$\theta = 0, \quad \left|1 - \mu/r^2\right| = \left|1 - \mu/r_0^2\right| e^{-2\mu t} \qquad [A.4]$$

If $\mu < 0$, when t increases, r decreases and tends toward zero for infinite t. The origin of the axes, therefore, is asymptotically stable, and the solution rolls up in a spiral around that origin.

If $\mu = 0$, the equation needing to be solved is: $\frac{dr}{dt} = -r^3$. This gives us: $r = 1/\sqrt{1/r_0^{\,2} + 2t}$, and in this system, r tends toward 0 as $t \to \infty$.

If $\mu > 0$, then $\left(r^2 - \mu\right)$ may be positive or negative. There is convergence toward $r \to \sqrt{\mu}$. This proves the stability of the curve $r = \mu^{1/2}$ in that zone, and the instability of the half-axis ($r = 0, \mu > 0$).

These results are shown in the stability diagram μ, r given in Figure A.1 and in the phase plane in Figures A.2 and A.3.

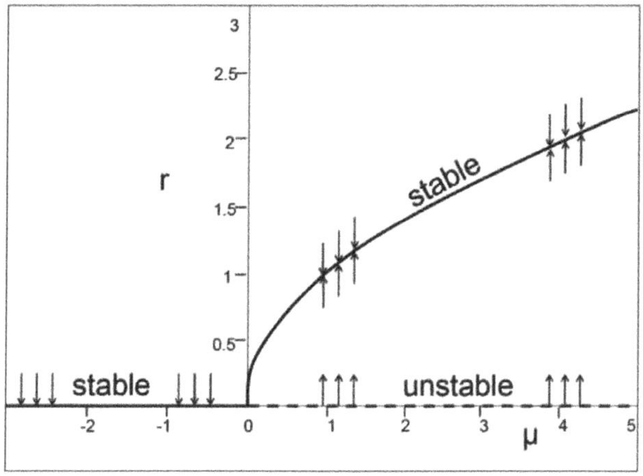

Figure A.1. *Diagram of the Hopf instability bifurcation, which shows the stability of the half-axis ($r = 0$, $\mu < 0$) and that of the curve ($r = \mu^{1/2}$, $\mu > 0$)*

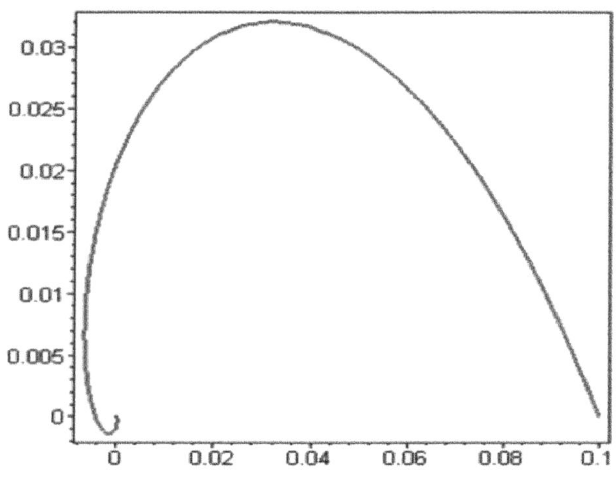

Figure A.2. *Trajectory converging toward the origin of the system of axes in the phase plane, obtained for a negative value of the control parameter: $\mu = -1$, $b = 101$, $r(\theta, b) = (b \exp 2\theta - 1)^{-1/2}$*

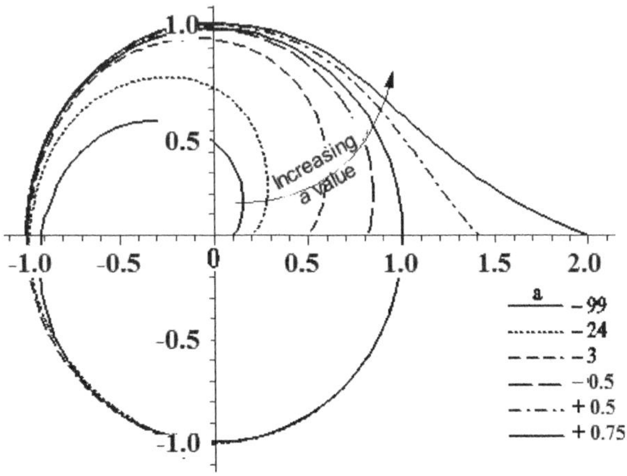

Figure A.3. *Trajectories in the phase plane converging toward the circle of radius $\sqrt{\mu}$, obtained for a positive value of the control parameter: $\mu = +1$; $a = (-99), (-24); (-3), (-0.5), (+0.5), (+0.75)$; $r(\theta, a) = [1 - a \exp(2\theta - 1)]^{-1/2}$*

Bibliography

[ABR 89] ABRAMZON B., SIRIGNANO W.A., "Droplet vaporization model for spray combustion calculations", *International Journal of Heat and Mass Transfer*, vol. 32, no. 9, pp. 1605–1618, 1989.

[AGB 11] AGBAGLAH G., DELAUCX S., FUSTER D., *et al.*, "Parallel simulation of multiphase flows using octree adaptativity and the volume-of-fluid method", *Comptes Rendus de Mécanique,* vol. 339, nos. 2–3, pp. 194–207, 2011.

[ALL 14] ALLOUCHE R., HAOUI R., RENANE R., "Numerical simulation of reactive flow in non-equilibrium behind a strong shock wave during re-entry into earth's atmosphere", *Mechanics & Industry*, vol. 15, no. 1, pp. 81–87, 2014.

[ATZ 06] ATZLER F., DEMOULIN F.X., LAWES M., *et al.*, "Burning rates and flame oscillations in globally homogeneous two-phase mixtures (flame speed oscillations in droplet cloud flames)", *Combustion Science and Technology*, vol. 178, no. 12, pp. 2177–2198, 2006.

[BAB 02] BABINSKY E., SOJSKA P.E., "Modeling drop size distributions", *Progress in Energy and Combustion Science*, vol. 28, no. 10, pp. 303–329, 2002.

[BAR 73] BARRERE M., PRUD'HOMME R., *Equations fondamentales de l'Aérothermochimie*, Masson et Cie, Paris, 1973.

[BEA 06] BEAU P.-A., Modélisation de l'atomisation d'un jet liquide. Application aux sprays Diesel, Doctoral Thesis, CoRIA-University of Rouen, 19 December 2006.

[BEN 54] BENJAMIN T.B., URSELL F., "The stability of the plane free surface of a liquid in vertical periodic motion", *Proceedings of the Royal Society of London*, vol. 255, pp. 505–515, 1954.

[BLO 03] BLOKKEEL G., MURA A., DEMOULIN F.X., *et al.*, "A continuous modeling approach for describing the atomization process from inside the injector to the final spray", *ICLASS*, Sorrento, Italy, 2003.

[BOC 94] BOCKHORN H., "A short introduction", *Soot Formation in Combustion – Mechanisms and Models*, Springer, Berlin, pp. 3–8, 1994.

[BOR 95] BORGHI R., LACAS F., "Modeling of liquid propellant spray combustion in rocket engine combustor", *Second International Symposium on Liquid Rocket Propulsion*, ONERA-Châtillon, 19-21 June 1995.

[BOU 09] BOUKRA M., CARTELLIER A., DUCASSE E., *et al.*, "Use of Faraday instabilities to enhance fuel pulverisation in air-blast Atomisers", *Comptes Rendus Mécanique*, vol. 337, pp. 492–503, 2009.

[BOZ 04] BOZIER O., Contribution à l'étude des caractéristiques de combustion isochore d'une suspension de particules solides réactives. Génération de la suspension; influence de l'état initial du mélange, Doctoral Thesis, ENSMA-Poitiers, 14 September 2004.

[BOZ 10] BOZIER O., VEYSSIÈRE B., "Study of the mechanisms of dust suspension generation in a closed vessel under microgravity conditions", *Microgravity Science and Technology*, vol. 22, pp. 233–248, 2010.

[BRE 07] BREMOND N., CLANET C., VILLERMAUX E., "Atomization of undulating liquid sheets", *Journal of Fluid Mechanics*, vol. 585, pp. 421–456, 2007.

[BR0 03] BROWN P.P., LAWLER D.F., "Sphere drag and settling velocity revisited", *Journal of Environmental Engineering*, vol. 129, pp. 222–231, 2003.

[CAR 64] CARLSON D.J., HOGLUND R.F., "Particle drag and heat transfer in rocket nozzles", *AIAA Journal*, vol. 2, no. 11, pp. 1980–1984, 1964.

[CHA 13] CHAHINE G., "Bubble Dynamics", *Research and Development in Applied Sciences*. Available at: http://www.dynaflow-inc.com/Publications/FullTextReports/Fulltext_Bubble_Cav.htm, 2013.

[CHE 97] CHEN Y.-J., STEEN P.H., "Dynamics of inviscid capillary breakup: collapse and pinchoff of a film bridge", *Journal of Fluid Mechanics*, vol. 341, 1997.

[CLI 78] CLIFT R., GRACE J.R., WEBER M.E., *Bubbles, Drops and Particles*, Academic, New York, pp. 306–314, 1978.

[COE 07] COELHO P.J., "Numerical simulation of the interaction between turbulence and radiation in reactive flows", *Progress in Energy and Combustion Science*, vol. 33, pp. 311–383, 2007.

[COR 56] CORSIN S., LUMLEY J., "On the equation of motion for a particle in a turbulent fluid", *Applied Science Research A*, vol. 6, pp. 114–116, 1956.

[COU 96a] COUSIN J., Prédiction des distributions granulométriques par le formalisme d'entropie maximum. Application à plusieurs injecteurs mécaniques, Doctoral Thesis, University of Rouen, 28 October 1996.

[COU 96b] COUSIN J., DUMOUCHEL C. "Effect of viscosity on the linear instability of a flat liquid sheet", *Atomisation and Spray*, vol. 6, pp. 563–576, 1996.

[CRI 00] CRIADO C., GALAN-MONTENEGRO P., VELASQUEZ P., *et al.*, "Diffusion with general conditions in electrochemical systems", *Journal of Electroanalytical Chemistry*, vol. 488, pp. 59–63, 2000.

[CRY 92] CRYER C., STEEN P.H., "Collapse of the soap-film bridge: Quasistatic description", *Journal of Colloid Interface Science*, vol. 154, pp. 276, 1992.

[DAS 04] DA SILVA M.L., Simulation des propriétés radiatives du plasma entourant un véhicule traversant une atmosphère planétaire à vitesse hypersonique – application à la planète Mars, Doctoral Thesis, University of Orléans, 2004.

[DAS 13] DA SILVA M.L., MARRAFFA L., "Radiative transfer in entry flows", *10th International Planetary Probe Workshop*, EDL Short Course, San José, California, United States, 15-21 June 2013.

[DEL 99] DELHAYE J.-M., ZUBER N., *Multiphase Science and Technology*, Hemisphere Publishing Corporation, 1999.

[DEM 12] DEMARCO BULL R.A., Modeling thermal radiation and soot formation in buoyant diffusion flames, Doctoral Thesis, University of Aix-Marseille, 2012.

[DEM 06] DEMOULIN F.-X., Contribution à la modélisation de la combustion turbulente des milieux à deux phases, Doctoral Thesis, CORIA, University of Rouen, 2006.

[DES 08] DESHMUKH K.V., MODEST M.F., HAWORTH D.C., "Direct numerical simulation of turbulence-radiation interactions in a statistically one-dimensional nonpremixed system", *Journal of Quantitative Spectroscopy and Radiative Transfer*, vol. 109, pp. 2391–2400, 2008.

[DIM 86] DIMOTAKIS P.E., "Two-dimensional shear layer entrainment", *AIAA Journal*, vol. 24, no. 11, pp. 1791–1796, 1986.

[DOD 95] DODEMAND E., PRUD'HOMME R., KUENTZMANN P., "Influence of unsteady forces acting on a particle in a suspension. Application to the sound propagation", *International Journal of Multiphase Flow*, vol. 21, no. 1, pp. 27–51, 1995.

[DOM 54] DOMBROWSKI N., FRASER R.P., "A photographic investigation into the disintegration of liquid sheets", *Philosophical Transactions of the Royal Society of London A*, vol. 247, pp. 101–130, 1954.

[DOM 62] DOMBROWSKI N., HOOPER P.C., "The effect of ambient density on drop formation in sprays", *Chemical Engineering Science*, vol. 17, pp. 291–305, 1962.

[DOM 63] DOMBROWSKI N., JOHNS W.R., "The aerodynamic instability and disintegration of viscous liquid sheets", *Chemical Engineering Science*, vol. 18, pp. 203–214, 1963.

[DOM 82] DOMBROWSKI N., "On possibility of determination of disperse composition of two-phase flow using light scattering under small angles", *Teplofizika Vysokich Temperatur*, 1982.

[DOR 89] DORDAIN J.-J., PRUD'HOMME R., BOUTTES J.F., Fluides hétérogènes, Majeure Sciences de l'espace, Ecole polytechnique, 1988-1989.

[DOR 12] DOREY L.-H., Modélisation des phénomènes couplés combustion – formation des suies – transferts radiatifs dans les chambres de combustion de turbine à gaz, Doctoral Thesis, ECP-ONERA, Paris, 1 June 2012.

[DUB 06] DUBOIS J., SMITH A.J., "ESA PlasmA RAdiation DatabasE (PARADE). Development history, status, current, developments and future prospects", *The 2nd International Workshop on Radiation of High Temperature Gases in Atmospheric Entry*, Università La Sapienza, Rome, Italy, 6–8 September 2006.

[DUD 76] DUDECK M., Equations linéarisées à l'interface, Rapport du Laboratoire d'aérothermique du CNRS, nos. 76–1, 1976.

[DUP 00] DUPAYS J., FABIGNON Y., ORLANDI O., *et al.*, "Combustion of aluminium particles in solid rocket motors", *ODAS 2000 – ONERA/DLR Aerospace Symposium*, Berlin, Germany, 15-16 June 2000.

[DUP 06] DUPOIRIEUX F., TESSÉ L., AVILA S., *et al.*, "An optimized reciprocity Monte Carlo method for the calculation of radiative transfer in media of various optical thicknesses", *International Journal of Heat and Mass Transfer*, vol. 49, pp. 1310–1319, 2006.

[DUV 02] DUVAL R., Transferts radiatifs dans les moteurs à propergol solide, Doctoral Thesis, Ecole centrale de Paris, 2002.

[EGG 97] EGGERS J., "Nonlinear dynamics and breakup of free-surface flows", *Reviews of Modern Physics*, vol. 69, p. 865, 1997.

[EMM 56] EMMONS H.W., "The film combustion of liquid fuel", *Zeitschrift für Angewandte Mathematik und Mechanik (ZAMM)*, vol. 36, pp. 60–71, 1956.

[FAB 92] FABRE J., LINÉ A., "Modeling of two phase slug flow", *Annu. Rev. Fluid Mech.*, vol. 24, pp. 21–46, 1992.

[FER 84] FERNANDEZ-PELLO A.C., "Flame spread modeling", *Combustion Science and Tecnology*, vol. 39, pp. 119–134, 1984.

[FER 09] FERNANDEZ V.G., BERTHOUMIEU P., LAVERGNE G., "Primary Atomization in Water and Kerosene Liquid Sheets At High Pressure", *ICLASS 2009, 11th Triennial International Annual Conference on Liquid Atomization and Spray Systems*, Vail, Colorado, United States, July 2009.

[FLA 90] FLAMENT C., "Chemical and vibrational non-equilibrium nozzle flow calculation by an implicit upwind method", *Proceedings of the 8th GAMM-Conference, Notes on Num. Fluid Mech.*, vol. 29, pp. 128–137, 1990.

[FOR 67] FORTIER A., *Mécanique des suspensions*, Masson, Paris, 1967.

[FRA 07] FRACKOVIAK B., Approche expérimentale et simulation numérique des effets d'interactions entre gouttes en évaporation, Doctoral Thesis, ENSAE-ONERA, Toulouse, 26 janvier 2007.

[FRE 89] FRENKLACH M., "On the driving force of PAH production", *Symposium (International) on Combustion*, vol. 22, no. 1, pp. 1075–1082, 1989.

[FRE 94] FRENKLACH M., WANG H., "Detailed mechanism and modeling of soot particle formation", in BOCKHORN H. (ed.), *Soot Formation in Combustion: Mechanisms and Models*, pp. 165–192, Springer-Verlag, Berlin, 1994.

[GAT 83] GATIGNOL R., "The Faxén formulae for a rigid sphere in an unsteady non-uniform Stokes flow", *Journal de Mécanique Théorique et Appliquée*, vol. 1, pp. 143–160, 1983.

[GAT 01] GATIGNOL R., PRUD'HOMME R., *Mechanical and thermodynamical modeling of fluid interfaces*, Series on Advances in Mathematics for Applied Sciences, World Scientific, vol. 58, 2001.

[GER 62] GERMAIN P., *Mécanique des milieux continus*, Masson, Paris, 1962.

[GLA 61] GLAUZ R.D., "Combined subsonic supersonic gas particle flow", *ARS Preprint*, nos. 1717–61, 1961.

[GOL 29] GOLDSTEIN S., "The steady flow of viscous fluid past a fixed spherical obstacle at small Reynolds number", *Proceedings of the Royal Society A*, vol. 123, pp. 225–235, 1929.

[GOO 97] GOODRIDGE C.L., TAO SHI W., HENTSCHEL H.G.E., *et al.*, "Viscous effects in droplet ejecting capillary waves", *Physical Review E*, vol. 56, pp. 472–475, 1997.

[GOR 07] GOROKHOVSKI M., HERRMANN M., "Modeling primary atomization", *Annual Review of Fluid Mechanics*, vol. 40, pp. 343–366, 2007.

[GOR 03] GOROKHOVSKI M., SAVELIEV V., "Analyses of Kolmogorov's model of breakup", *Physics of Fluids*, vol. 15, pp. 184–192, 2003.

[GRO 63] DE GROOT S.R., MAZUR P., *Non Equilibrium Thermodynamics*, North-Holland Publishing Company, Amsterdam, 1963.

[GRO 69] DE GROOT S.R., MAZUR P., *Non Equilibrium Thermodynamics*, North-Holland Publishing Company, Amsterdam, 1969.

[GUP 90] GUPTA R.N., YOSS J.M., THOMPSON R.A., *et al.*, "A review of reaction rates and thermodynamic and transport properties for an 11-species air model for chemical and thermal nonequilibrium calculations to 30000 K", *NASA-RP-1232*, 1990.

[GUY 13] GUY A., Collisional-radiative and macroscopic models for the thermochemical relaxation of non-equilibrium hypersonic flows, Doctoral Thesis, Ecole Centrale de Paris, 16 December 2013.

[HAN 63] HANSON A.R., DOMICH G.E., ADAMS H.S., "Shock tube investigation of the break-up of drops by air blasts", *Physics of fluids*, vol. 6, pp. 1070–1080, 1963.

[HIN 75] HINZE J.O., *Turbulence*, 2nd edition, McGraw-Hill, New York, pp. 460–471, 1975.

[HOM 10] HOMANN H., BEC J., "Finite-size effects in the dynamics of neutrally buoyant particles in turbulent flow", *Journal of Fluid Mechanics*, vol. 651, pp. 81–91, 2010.

[HUL 65] HULTBERG J.A., SOO S.L., "Flow of a gas solid suspension through a nozzle", *AIAA Paper*, nos. 65–6, 1965.

[JEN 04] JENNY M., DUSEK J., BOUCHET G., "Instabilities and transition of a sphere falling or ascending freely in a Newtonian fluid", *Journal of Fluid Mechanics*, vol. 508, pp. 201–239, 2004.

[JOU 93] JOULAIN P., "Diffusion Flame Combustion in Oxidizing Flow", *Micro-Gravity Science and Technology*, vol. 4, pp. 178–191, 1993.

[JOU 01] JOUMANI Y., Transferts radiatifs dans les moteurs à propergol solide, Doctoral Thesis, University of Valenciennes and Hainaut Cambraisis, 2001.

[KAM 13] KAMINSKA A., SZYMANSKI Z., HOFFMAN J., *et al.*, "Ecoulements ionisés dans les plasmatrons pour la simulation des conditions d'entrée en atmosphères planétaires", in IZARRA C., VALLEE M. (eds.), *Modélisation: atomes, molécules, plasmas, systèmes dynamiques*, Presses Universitaires, Bourges, 23 and 24 May 2013.

[KLI 60] KLIEGEL J.R., "One dimensional flow of a gas particle system", *IAS Paper*, nos. 60–5, 1960.

[KRA 01] KRALCHEVSKY P.A., NAGAYAMA K., *Particles at fluid interfaces and membranes*, Elsevier, Amsterdam, 2001.

[KRÜ 90] KRÜLLE G., MAYER W., SCHLEY C.A., "Recent advances in H_2/O_2 high pressure coaxial injector performance analysis", *AIAA/SAE/ASME/ASEE 26th Joint Propulsion Conference*, Orlando, United States, 16-18 July 1990.

[KUE 73] KUENTZMANN P., *Aérothermochimie des suspensions*, Mémoire des sciences physiques, Fasc. 72, Gauthiers-Villars, Paris, 1973.

[LAL 06] LALO M., Atomisation d'un film mince liquide par action combinée des instabilités de Kelvin-Helmholtz et de Faraday. Application aux injecteurs aérodynamiques des turbomachines aéronautiques, Doctoral Thesis, ENSAE-ONERA, Toulouse, 21 December 2006.

[LAM 45] LAMB H., *Hydrodynamic*, Cambridge University Press, 1945.

[LAM 09] LAMET J.-M., Transferts radiatifs dans les écoulements hypersoniques de rentrée atmosphérique terrestre, Doctoral Thesis, Ecole Centrale de Paris, 2009.

[LAN 71] LANDAU L., LIFSCHITZ E., *Mécanique des fluides*, MIR, Moscow, 1971.

[LAN 62] LANG R.J., "Ultrasonic atomisation of liquids", *The Journal of the Acoustical Society of America*, vol. 34, no. 1, 1962.

[LAS 98] LASHERAS J.C., VILLERMAUX E., HOPFINGER E., "Break-up and atomisation of a round water jet by a high-spead annular jet", *Journal of Fluid Mechanics*, vol. 357, pp. 351–379, 1998.

[LAW 77] LAW C.K., SIRIGNANO W.A., "Unsteady droplet combustion with droplet heating-II: conduction limit", *Combustion and Flame*, vol. 28, pp. 17–186, 1977.

[LEG 99] LEGENDRE D., COLIN C., FABRE J., *et al.*, "Influence of gravity upon the bubble distribution in a turbulent pipe flow: comparison between numerical simulations and experimental data", *Journal de Chimie Physique*, vol. 96, pp. 951–957, 1999.

[LEG 05] LEGROS G., FUENTES A., BAILLARGEAT J., *et al.*, "Three-dimensional recomposition of the absorption field inside a non-buoyant sooting diffusion flame", *Optical Letters*, vol. 30, pp. 3311–3313, 2005.

[LEG 06] LEGROS G., JOULAIN P., VALENTON J.P., *et al.*, "Soot volume fraction measurements in a three-dimensional laminar diffusion flame established in microgravity", *Combustion Science and Technology*, vol. 178, pp. 813–835, 2006.

[LEG 09] LEGROS G., FUENTES A., ROUVREAU S., *et al.*, "Transport mechanisms controlling soot production inside a non-buoyant laminar diffusion flame", *Proceedings of the Combustion Institute*, vol. 32, pp. 2461–2470, 2009.

[LEM 10] LE MOYNE L., "Atomisation, pulvérisation et aérosols", *Techniques de l'Ingénieur*, Référence AF 3622, 2010.

[LHU 11] LHUISSIER H., Nappes, trous, ligaments et gouttes, Doctoral Thesis, University of Provence-Aix-Marseille I, 2011.

[LIG 07] LIGNELL D.O., CHEN J.H., SMITH P.J., *et al.*, "The effect of flame structure on soot formation and transport in turbulent non-premixed flames using direct numerical simulation", *Combustion and Flame*, vol. 151, pp. 2–28, 2007.

[LUD 73] LUDWIG C.B., MALKMUS W., REARDON J.E., *et al.*, Handbook of infrared radiation from combustion gases, NASA Technical Report SP-3080, Washington, 1973.

[LYP 61] LYPE E.F., "One dimensional analysis of non-isentropic two phase flow", *ARS Preprint*, nos. 1605–61, 1961.

[MAC 14] MACHICOANE N., ZIMMERMANN R., FIABANE L., *et al.*, "Large sphere motion in a nonhomogeneous turbulent flow", *New Journal of Physics*, vol. 16, 2014.

[MAR 70] MARBLE F.E., "Dynamics of dusty gases", *Ann. Rev. Fluid Mech*, vol. 2, pp. 397–446, 1970.

[MAR 77] MARBLE F.E., BROADWELL J.E., The coherent flame model for turbulent chemical reactions, Technical Report TRW-9-PU, Project Squid Headquarters, Purdue University, 1977.

[MAR 01] MARMOTTANT P., Atomisation d'un courant liquide dans un courant gazeux, Doctoral Thesis, Institut polytechnique de Grenoble, 2001.

[MAR 04a] MARMOTTANT P., VILLERMAUX E., "On spray formation", *Journal of Fluid Mechanics*, vol. 498, pp. 73–111, 2004.

[MAR 04b] MARMOTTANT P., VILLERMAUX E., "Fragmentation of stretched liquid ligaments", *Physics of fluids*, vol. 16, no. 8, pp. 2732–2741, 2004.

[MAU 08] MAURIOT Y., PRUD'HOMME R., High frequency combustion instability activities – Droplet dynamics (Task 3.2), Rapport Technique ONERA, no. 4/12499 DEFA, November 2008.

[MAU 14] MAURIOT Y., PRUD'HOMME R., Assessment of evaporation equilibrium and stability concerning an acoustically excited drop in combustion products, *Comptes Rendus Mécanique*, vol. 342, no. 4, pp. 240–253, 2014.

[MAX 83] MAXEY M.R., RILEY J.J., "Equation of motion for a small rigid sphere in a uniform flow", *Physics of Fluids*, vol. 26, no. 4, pp. 883–889, 1983.

[MEH 10] MEHTA R.S., HAWORH D.C., MODEST M.F., "Composition PDF/photon Monte Carlo modeling of moderately sooting turbulent jet flames", *Combustion and Flame*, vol. 157, no. 4, pp. 982–994, 2010.

[MOD 03] MODEST M.F., *Radiative Heat Transfer*, 2nd edition, Academic Press, New York, 2003.

[MUG 51] MUGELE R.A., EVANS H.I., "Droplet size distribution in sprays", *Ind. Engng. Chem*, vol. 43, no. 4, pp. 1317–24, 1951.

[MUR 06] MURRONE A., ECLIPPS. Adaptation de la chaîne de calcul SPIREE à la problématique du givrage, Rapport Technique ONERA, no. 7, February 2006.

[NIC 62] NICKLING D.J., WILKES J.O., DAVIDSON J.F., "Two phase flow in vertical tubes", *Transactions of the Chemical Institute of Chemical Engineering*, vol. 40, pp. 61–68, 1962.

[NIC 05] NICOLI C., HALDENWANG P., SUARD S., "Analysis of pulsating spray-flames propagating in lean two-phase mixtures with unity Lewis number", *Combustion and flame*, vol. 143, pp. 299–312, 2005.

[NIC 09a] NICOLI C., HALDENWANG P., "Flamme de brouillard en présence d'une onde acoustique", *Colloque du GDR Micropesanteur Fondamentale et Appliquée*, La Villa Clythia – Fréjus, 26, 27 and 28 November 2009.

[NIC 09b] NICOLI C., HALDENWANG P., "Etude en boucle ouverte de l'action résonnante d'une onde acoustique sur une flamme de brouillard", *19e Congrès Français de Mécanique*, Marseille, 24-28 August 2009.

[NIC 10] NICOLI C., HALDENWANG P., "A resonant response of self-pulsating spray-flame submitted to acoustic wave", *Combustion Science and Technology*, vol. 182, nos. 4–6, 2010.

[NIG 91] NIGMATULIN R.L., *Dynamics of Multiphase Media*, Hemisphere Publishing Corporation, 1991.

[OLE 97] OLESEN M.J., Prediction of drop-size distributions based on ligament breakup, Doctoral Thesis, Queen's University, Kingston, Canada, July 1997.

[ORO 97] ORON A., DAVIS S.H., BANKOFF S.G., "Long-scale evolution of thin liquid films", *Review of Modern Physics*, vol. 69, no. 3, 1997.

[PAD 73] PADDAY J., PITT A., "The stability of axisymmetric menisci", *Philosophical Transactions of the Royal Society of London A*, vol. 275, pp. 489–528, 1973.

[PAD 97] PADDAY J., PÉTRÉ J., RUSU C., *et al.*, "The shape stability and breakage of pendant liquid bridges", *Journal of Fluid Mechanics*, vol. 352, pp. 177–204, 1997.

[PAR 85] PARK C., "On convergence of computation of reacting flows", *AIAA paper*, 1985.

[PAR 89] PARK C., "Assessment of Two-Temperature Kinetic Model for Ionizing Air", *Journal of Thermophysics*, July 1989.

[PAR 08] PARKER S.H., SUH H.K., LEE C.S., "Effect of cavitating flow on the flow and fuel atomisation characteristics of biodiesel and diesel fuels", *Energy and Fuels*, vol. 22, pp. 605–613, 2008.

[PAS 06] PASCAUD S., Vers la simulation aux grandes échelles des écoulements turbulents diphasiques réactifs: application aux foyers aéronautiques, Doctoral Thesis, Institut national polytechnique de Toulouse, 28 June 2006.

[PET 00] PETERS N., *Turbulent combustion*, Cambridge University Press, 2000.

[PLA 12] PLANCHETTE C., LAURENCEAU E., BRENN G., "The onset of fragmentation in binary liquid drop collisions", *Journal of Fluid Mechanics*, vol. 702, pp. 7–25, 2012.

[POI 05] POINSSOT T., VEYNANTE D., *Theoretical and Numerical Combustion*, 2nd edition, Edwards, Flourtown, 2005.

[PRO 57] PROUDMAN I., PEARSON J.R.A., "Expansions at small Reynolds numbers for the flow past a sphere and a circular cylinder", *Journal of Fluid Mechanics*, vol. 2, pp. 237–262, 1957.

[PRU 96] PRUD'HOMME R., JAMET N., Notions de base sur les équations des brouillards, ONERA technical report, June 1996.

[PRU 98] PRUD'HOMME R., La méthode du maximum d'entropie appliqué à l'atomisation des jets liquides, ONERA technical report, 1998.

[PRU 03] PRUD'HOMME R., HABIBALLAH M., NICOLE A., "On Transient heating and dynamic response to acoustic oscillation of a vaporizing droplet", *XXXI Summer School – Conference "Advanced Problems in Mechanic" (APM2003)*, Saint-Petersburg, Russia, 23 July 2003.

[PRU 04] PRUD'HOMME R., PRÉAU S., "Echanges thermiques et phases fluides", *ICTEA-HT3-06*, Beirut, Lebanon, 31 May to 4 June 2004.

[PRU 08] PRUD'HOMME R., Analyse théorique du chauffage transitoire et de la réponse dynamique à une oscillation acoustique d'une goutte en évaporation, ONERA final report, June 2008.

[PRU 10a] PRUD'HOMME R., *Flows of Reactive Fluids*, FMIA Series, vol. 94, Springer, Berlin, 2010.

[PRU 10b] PRUD'HOMME R., HABIBALLAH M., MATUSZEWSKI L., *et al.*, "Theoretical analysis of dynamic response of a vaporizing droplet to an acoustic oscillation", *Journal of Propulsion and Power*, vol. 26, no. 1, pp. 74–83, 2010.

[PRU 12] PRUD'HOMME R., *Flows of Reactive Fluids*, ISTE, London, and John Wiley & Sons, New York, 2012.

[PRU 13] PRUD'HOMME R., *Flows of reactive Fluids in Homogeneous Mixtures*, ISTE, London, and John Wiley & Sons, New York, 2013.

[QUR 07] QURESHI N.M., BOURGOIN M., BAUDET C., *et al.*, "Turbulent tranport of materials particles: an experimental investigation of finite size effects", *Physical Review Letters*, vol. 99, p. 184502, 2007.

[RAN 08] RANGWALA A.S., "Flame Spread Analysis using a Variable B-Number", *Fire Safety Science–proceedings of the ninth International Symposium*, pp. 243–254, 2008.

[RAY 78] RAYLEIGH J.W.S., "On the instability of jets", *Proceedings of the London Mathematical Society*, vol. 1–10, no. 1, pp. 4–13, 1878.

[RAY 97] RAYNAL L., Instabilité et entraînement à l'interface d'une couche de mélange liquide gaz, Doctoral Thesis, Joseph Fourier University (Grenoble I), 7 October 1997.

[REF 11] REFLOCH A., COURBET B., MURRONE A., *et al.*, "CEDRE Software", *Aerospace Laboratory*, no. 2, March 2011.

[RIT 13] RITTER H., MARRAFFA L., FERRACINA L., *et al.*, "ESA technology roadmaps ongoing and planned developments", *7th European Workshop on Thermal Protection Systems & Hot Structure*, 8 April 2013.

[ROG 10] ROGER M., COELHO P.J., DA SILVA C.B., "The influence of the non-resolved scales on thermal radiation in large eddy simulation of turbulent flows: a fundamental study", *International Journal of Heat and Mass Transfer*, vol. 53, pp. 2897–2907, 2010.

[SAG 06] SAGAUT P., *Large Eddy Simulation for Incompressible Flows*, Springer, Berlin, 2006.

[SEL 98] SELLENS R., ZENG W., "Modeling Spectral Influences in Droplet/Turbulence Interaction", *ILASS Americas, 11th Annual Conference on Liquid Atomization and Spray Systems*, Sacramento, United States, May 1998.

[SEN 09] SENONER J.M., SANJOSÉ M., LEDERLIN T., *et al.*, "Eulerian and Lagrangian Large-Eddy Simulations of an evaporating two-phase flow", *Comptes Rendus de Mécanique*, vol. 337, pp. 458–468, 2009.

[SGR 07] SGRO L.A., FILIPPO A.D., LANZUOLO G., *et al.*, "Characterization of nanoparticles of organic carbon (NOC) produced in rich premixed flames by differential mobility analysis", *Proceedings of the Combustion Institute*, vol. 31, no. 1, pp. 631–638, 2007.

[SIN 95] SINDAYIHEBURA D., Pulvérisation ultrasonique de films liquides de faible épaisseur, Doctoral Thesis, Université Catholique de Louvain, 1995.

[SIT 85] SITU M., SCHETZ J.A., "Fragmentation Computational and experimental study of the effect of mass transfer on liquid-jet breakup", *AIAA Journal*, vol. 23, pp. 254–261, 1985.

[SOO 90] SOO S.L., *Multiphase Fluid Dynamics*, Science Press and Gower Technical, 1990.

[SOU 06] SOUBRIE T., Prise en compte de l'ionisation et du rayonnement dans la modélisation des écoulements de rentrée terrestre ou martienne, Doctoral Thesis, Ecole nationale supérieure de l'aéronautique et de l'espace, 2006.

[SUN 11] SUNDERLAND P.B., QUINTIERE J.G., TABAKA G.A., *et al.*, "Analysis and measurement of candle flame shapes", *Proceedings of the Combustion Institute*, vol. 33, pp. 2489–2496, 2011.

[TAK 09] TAKAHASHI F., "Candle and jet diffusion flames: mechanisms of combustion under gravity and microgravity conditions", in JAROSINSKI J., VEYSSIÈRE B., (eds.), *Combustion Phenomena*, CRC Press, Boca Raton, pp. 170–178, 2009.

[TCH 47] TCHEN C.M., Mean value and correlation problems connected with the motion of small particles suspended in a turbulent fluid, Doctoral Thesis, University of Delft, 1947.

[TES 01] TESSÉ L., Modélisation des transferts radiatifs dans les flammes turbulentes par une méthode de Monte Carlo, Doctoral Thesis, Ecole Centrale de Paris, 2001.

[TES 02] TESSÉ L., DUPOIRIEUX F., ZAMUNER B., *et al.*, "Radiative transfer in real gases using reciprocal and forward Monte Carlo methods and a correlated-k approach", *International Journal of Heat and Mass Transfer*, vol. 45, pp. 2797–2814, 2002.

[TES 04] TESSÉ L., DUPOIRIEUX F., TAINE J., "Monte Carlo modeling of radiative transfer in in a turbulent sooty flame", *International Journal of Heat and Mass Transfer*, vol. 47, pp. 555–572, 2004.

[TES 11] TESSÉ L., LAMET J.-M., "Radiative Transfer Modeling Developed at ONERA for Numerical Simulations of Reactive Flows", *Journal Aerospace Lab*, no. 2, pp. 1–19, 2011.

[THE 08] THEOFANOUS T.G., LI G.J., "On the physics of aerobreakup", *Physics of Fluids*, vol. 20, pp. 1070–1080, 2008.

[TIO 93] TIO K.-K., LINIAN A., LASHERAS J.C., *et al*, "On the dynamics of buoyant and heavy particles in a periodic Stuart vortex flow", *Journal of Fluid Mechanics*, vol. 254, pp. 671–699, 1993.

[TOR 01] TORERO J., VIETORIS T., LEGROS G., *et al.*, "Evaluation d'un nombre de transfert de masse réel d'une flamme ascendante", *Journal de Physique IV France*, vol. 11, 2001.

[TOR 02] TORERO J., VIETORIS T., LEGROS G., *et al.*, "Estimation of a Total Mass Transfer Number from the Stand-off Distance of a Spreading Flame", *Combustion Science and Technology*, vol. 174, pp. 187–203, 2002.

[TOT 05] TÓTH B., LEMA M.R., STEELANT J., *et al.*, "Assessment of slag accumulation in solid rocket boosters: Part I, single-phase", *AIAA Paper*, pp. 1–10, 2005.

[TRO 94] TROUVÉ A., POINSOT T., "The evolution equation for the flame surface density in premixed combustion", *Journal of fluid mechanics*, vol. 278, pp. 1–31, 1994.

[TRY 11] TRYGGVASON G., SCARDOVELLI R., ZALESKI S., *Direct numerical simulations of gas-liquid multiphase flows*, Cambridge University Press, 2011.

[UHL 05] UHLMANN M., "An immersed boundary method with direct forcing for the simulation of particulate flows", *Journal of Computational Physics*, vol. 209, pp. 448–476, 2005.

[VAL 01] VALLET A., BURLUKA A.A., BORGHI R., "Development of an Eulerian model for the atomization of a liquid jet", *Atomization and Sprays*, vol. 11, pp. 619–642, 2001.

[VEY 03] VEYSSIERE B., BOZIER P., Mécanismes de propagation des flammes dans les mélanges hétérogènes gaz-particules solides, Bilan à 4 ans du GDR 2258 CNES-CNRS : Phénomènes de transport et transitions de Phase en Micropesanteur, Paris, September 2003.

[VIE 99] VIETORIS T., JOULAIN P., TORERO J., "Experimental characterisation of a laminar diffusion flame in micro-gravity", *Journal de Chimie Physique*, vol. 96, pp. 1022–1030, 1999.

[VIL 04] VILLERMAUX E., MARMOTTANT E., "Fragmentation of stretched liquid ligaments", *Physics of Fluids*, vol. 16, pp. 2732–2741, 2004.

[VIL 07] VILLERMAUX E., "Fragmentation", *Annual Review of Fluid Mechanics*, vol. 39, pp. 419–446, 2007.

[VIL 09] VILLERMAUX E., BOSSA B., "Single-drop fragmentation determine size distribution of raindrops", *Nature Physics*, vol. 5, pp. 696–702, September 2009.

[WAL 69] WALLIS G.B., *One-Dimensional Two-Phase Flow*, McGraw-Hill, New York, 1969.

[WAN 96] WANG H., DU D.X., SUNG C.J., *et al.*, "Experiments and numerical simulation on soot formation in opposed-jet ethylene diffusion flames", *Symposium (International) on Combustion*, vol. 26, no. 2, pp. 2359–2368, 1996.

[WAN 01] WANG H.Y., CORDEIRO P., JOULAIN P., "Direct numerical simulation of a laminar diffusion flame established over a horizontal flat plate in micro-gravity environment", *Journal de Physique IV France*, vol. 11, pp. 331–338, 2001.

[WAN 07] WANG H.Y., JOULAIN P., EL-RABII H., "Modélisation numérique et étude expérimentale sur la propagation d'une flamme de diffusion sous gravité réduite avec du transfert radiatif", *Colloque du GDR Micropesanteur Fondamentale et appliquée*, Fréjus, 26-28 November 2007.

[WIL 85] WILLIAMS F.A., *Combustion Theory*, 2nd edition, Benjamin Cummings, Menlo Park, 1985.

[XUE 03] XU F., EL–LEATHY M., KIM C.H., *et al.*, "Soot surface oxidation in hydrocarbon/air diffusion flames at atmospheric pressure", *Combustion and Flame*, vol. 132, nos. 1–2, pp. 43–57, 2003.

[YEC 05] YECKO P., ZALESKI S., "Transient growth in two-phase mixing layers", *Journal of Fluid Mechanics*, vol. 528, pp. 43–52, 2005.

[YOU 95] YOUNG J.B., "The fundamental equations of gas-droplet multiphase flow", *International Journal of Multiphase Flow*, vol. 21, no. 2, pp. 175–191, 1995.

[ZHA 05] ZAO B., YANG Z., LI Z., *et al.*, "Particle size distribution function of incipient soot in laminar premixed ethylene flames: effect of flame temperature", *Proceedings of the Combustion Institute*, vol. 30, no. 1, pp. 1641–1448, 2005.

Index

E, F

G, H, I

K, L, M

N, O, P

Q, R

S, T, V, W, Z

Other titles from ISTE in Fluid Mechanics

2014

PRUD'HOMME Roger
Flows and Chemical Reactions in an Electromagnetic Field

TARDU Sedat
Transport and Coherent Structures in Wall Turbulence

2013

BORGHI Roland, ANSELMET Fabien
Turbulent Multiphase Flows with Heat and Mass Transfer

DÉLERY Jean
Three-dimensional Separated Flow Topology: Critical Points, Separation Lines and Vortical Structures

PRUD'HOMME Roger
Flows and Chemical Reactions in Homogeneous Mixtures

2012

BOUTIER Alain
Laser Velocimetry in Fluid Mechanics

BOUTIER Alain
Laser Metrology in Fluid Mechanics: Granulometry, Temperature and Concentration Measurements

PRUD'HOMME Roger
Flows and Chemical Reactions

2011

MORY Mathieu
Fluid Mechanics for Chemical Engineering

TARDU Sedat
Statistical Approach to Wall Turbulence

2010

COLIN Stéphane
Microfluidics

DÉLERY Jean
Handbook of Compressible Aerodynamics

GUINOT Vincent
Waves Propagation in Fluids – 2nd edition

SOULI Mhamed, BENSON J. David
Arbitrary Lagrangian Eulerian and Fluid-structure Interaction

2009

FAVRE-MARINET Michel, TARDU Sedat
Convective Heat Transfer

2008

PEUBE Jean-Louis
Fundamentals of Fluid Mechanics and Transport Phenomena

SCHIESTEL Roland
Modeling and Simulation of Turbulent Flows

Printed and bound by CPI Group (UK) Ltd, Croydon, CR0 4YY

13/09/2026

14961016-0001